Shape of a Boy

For my three wonderful adventurers Josh, Ben and Freddie
And the best navigator a girl could wish for, Neil.

Shape of a Boy

Family life lessons in far-flung places

Kate Wickers

Aurum

First published in 2022 by Aurum,
an imprint of The Quarto Group.
The Old Brewery, 6 Blundell Street,
London, N7 9BH, United Kingdom.
www.QuartoKnows.com/Aurum

A catalogue record for this book is available from the British Library.

ISBN: 978-0-7112-6717-6
Ebook ISBN: 978-0-7112-6719-0
Audiobook ISBN: 978-0-7112-6720-6

1 2 3 4 5 6 7 8 9 10

Typeset in Kepler Std by SX Composing DTP, Rayleigh, Essex, SS6 7EF
Printed and bound by CPI Group (UK) Ltd, Croydon, CR0 4YY
Jacket Design by Hannah Naughton

Contents

Introduction

'Oh, but they won't remember any of it!' was the standard comment when I told family and friends of our latest travel plans to take our children to Borneo, Cambodia or Sri Lanka. 'Might as well just stick them in a cupboard until they're teenagers then,' was my usual tart reply.

My own childhood memories of holidays are of rock-pooling in Cornwall and caravanning in the Lake District, but there are a couple of others that stand out. On a road trip through France and Switzerland when I was five years old, we stayed in a run-down chateau, where by night we listened to the sound of furniture being dragged across the floor in the room above us, only to be told by the owner that the room was unoccupied. I made a friend of the owner's daughter, Mimi, who spoke no English but would call for me with a tuneful 'Coooeeeeee'. We drank chocolate milk out of glass bottles and tied ribbons around the tail of the chateau's cat. My parents ran out of money on the last day and we slept in our car in a field only to wake to the sight of my Dad staring down the barrel of an angry French farmer's shot gun. Another year, my Mum and I flew to Spain, while my Dad followed overland on his 1000cc motorbike. I remember the thrill, aged twelve, of riding pillion (without a crash helmet) along the esplanade in Barcelona, and the admiring glances my Dad's bike received from local teenage boys. I took my first solo trip at eighteen, booking into a small scruffy hotel in Paris near the Jardin de Tuileries, where I spent my time posing on benches pretending to read while smoking unfiltered French cigarettes. I realized then the possibilities travel gave to reinvent yourself, with no one there to say, 'That's not like you.'

In my second year of studying Media & Communications at university, a professor suggested that I should send some features out to publications with a view to getting published, giving this advice, 'Write about something you love'. It took me all of five seconds to decide that would be travel. I bought my first *Writers' and Artists' Yearbook* in 1994 and turning to the list of publications beginning with A, I sold my first feature on Prague to a magazine called *Active Life*. This was quickly followed by several commissions from *Adventure Travel*. I then went on to B (*Birdwatching*), C (*Cycling Today*) – you get the picture. By the time I graduated, I'd pro–gressed to T, writing regular adventure travel features for *The Telegraph*, covering stories such as *Tet* (New Year) in Vietnam, hiking the Annapurna Circuit in Nepal and trekking through Tiger Leaping Gorge in China.

My boyfriend (now husband), Neil, and I shared a love of travel, exploring Greece, Morocco and Turkey in the early days of our relationship, but soon striking out to India and Nepal, Ecuador and the Galapagos.

I didn't really consider what travelling with kids in tow would be like but when our first son, Josh, was born, I was determined that 'Have baby, will travel' would be my mantra. He was just three months old when we boarded a plane to Mallorca, at four months he came with us to Lisbon, at five months to Amsterdam. He couldn't have been a lovelier or easier companion. When Josh was nine months old, I found out that I was pregnant again. Surely two little travel buddies would be even better than one?

I was first and foremost a stay-at-home mum (pre-school childcare consisted of three afternoons at nursery for the boys while I wrote), taking writing assignments and editing projects on a wide variety of subjects from alternative medicine to horticulture to property, but travel writing was the work I loved the most. When Josh was three and Ben two, I decided it was time to venture further afield but an unexpected complication arose, when Josh was diagnosed with a life-threatening nut allergy. My instinct was to keep him in a nice safe bubble, never straying out of our comfort zone, however I didn't want Josh growing up with an allergy that held him back, too frightened to explore the world (and its glorious food). So, armed with an EpiPen we travelled and every country we visited I had the same terrifying but crystal-clear sentence translated into the local language: 'My son has a life-threatening allergy to nuts and nut oil'.

By the time our third son, Freddie, was born (when Josh was five and Ben three), we were confident travelling with kids, and felt sure that we'd continue to explore the world as a family of five (tuk-tuks, by the way, make great baby changing stations).

Shape of a Boy tells of the many quite-by-chance life lessons that our sons have learned while travelling. What we experienced as a family often took us by surprise. Most experiences were magical and exceptional, and on very rare occasions disconcerting. All were life-defining. I felt sure that these experiences would shape them into the adults they'd become, whether they remembered them or not. After all, memory is a tricky one to define, often triggered by a smell or taste; a photograph or story shared within a family. 'Remember the time when we . . . ?' Our conversations so often begin like this.

1

Israel & Jordan

A Lesson in Parenthood

with Ernie (Josh) in utero age fifteen weeks

I March 2000,
Jerusalem, Israel

*'Oh, Little town of Bethlehem, how noisy/crowded/commercial
we see thee lie'. I was glad to escape the scrum of tourists in
Manger Square, and head to the Chapel of the Milk Grotto
which, as excursions go, will probably rank as one of my
weirdest. Said to be the spot where Mary hid in an underground
cave to breastfeed Jesus as she and Joseph fled from King
Herod, there was a small queue of devotees waiting to pray to
'Our Lady of the Milk'. Feeling more than a little ridiculous
I joined them. So, the story goes that drops of Mary's milk fell
to the ground, turning the brown stones a creamy white. It's a
stretch for the imagination, even though there's 'evidence'
displayed – a piece of the original brown rock that Mary's
squirting breasts failed to hit. What's also unusual about the
place is that it is visited by both Catholic and Muslim women*

who are struggling to conceive or breastfeed. On the wall there's a notice that states that three thousand babies have been born to mothers who have prayed here. What it doesn't tell you is how many women have visited and remained childless. I'm guessing it's a lot more. Once in, I felt my cynicism ebb away, immediately charmed by the cosy atmosphere (would it be blasphemous to suggest that it would make a cracking bar?), with the uneven white walls and roof flickering with candlelight, softly illuminating the paintings of Mary and Jesus that decorate the walls. The altar was simple, adorned with a beautiful icon of Mary nursing Jesus, and there I found a young Arab woman kneeling in prayer. I hung back so as not to disturb her, wondering what her story was. When she rose, she looked back to me and smiled shyly, glancing at my stomach. I nodded and her smile widened. It was a lovely connection to make, and I felt immediately guilty for scoffing at her and the other women, who, most likely, were desperate to conceive. Who knows, if it had taken me longer to become pregnant perhaps I would have been on my knees beside her.

~

Neil was going on a hare-brained, team-building, camel-riding jolly in the Wadi Rum desert in Jordan (these were the days when large companies still had money to send employees on such extravagant non-senses), and I had a commission to write a feature on Jerusalem for *The Sunday Telegraph*. So, we thought we'd combine the two events with a few days in Israel together, before he departed for his desert adventure. The plan was that I'd stay on in Jerusalem to research my feature, then fly down to the coastal resort of Eilat, from where I would continue overland into Jordan, reconvening with Neil at the ancient city of Petra. I'd had my first scan and all was looking good, so there was absolutely no reason to take it easy at this stage of pregnancy (despite what my concerned family thought).

'I just wish you'd slow down a bit,' said my mum.

'In lots of places, you see pregnant women working in fields, and in India I even saw some on building sites carrying bricks,' I told her.

'Oh, those poor women.'

'My point is, as women we're designed to carry on our normal lives while we're pregnant, aren't we?'

'Depends what you call normal. Carrying bricks, though? I can't imagine it.'

I was beginning to wish I'd never mentioned the building site.

'I promise I won't do any bricklaying,' I told her. I really couldn't understand what she was worrying about. This wasn't a big deal . . . or so I thought.

Once Neil had left, I mooched around Jerusalem for a couple of days. Being alone never bothered me, especially in interesting cities and Jerusalem is without a doubt one of the most fascinating. Like walking through history in a living, breathing museum, I did what I normally did and wandered at whim. In Jerusalem's higgle-piggle arrangements of narrow streets, I knew that if I tried to follow a map, I'd end up confused and frustrated (I've never been the best navigator) and miss what was going on around me. Straying off course was the way I liked to travel, in the hope of discovering something unique, perhaps an artisan's workshop or a scene of local life, away from the tourist hotspots. As a travel writer, you're forever on the hunt for something new to write about, but on this day, halfway down one small street, I panicked. There wasn't a soul around. I looked back to where I'd just come from, suddenly unsure of which small lane had brought me to this spot. Later, I realized that I was experiencing pregnancy-induced brain fog, which had also caused me to put my passport in the minibar instead of the room safe.

Up ahead, a door swung open and a young Muslim woman stepped out, tugging at the hand of a small boy. She paused to lock the door, while her son kicked a stone around, his big brown eyes glancing shyly back to me, and seeming to ask, 'What are you doing here?' It was a good question and feeling safer in their company I stuck close, trusting that at some point they would lead me out of this maze and into a busier part of the old city. We parted ways, with shy smiles, at the Hurva Synagogue, and with reassuring landmarks again now visible, I headed up to the Western Wall, more commonly known as the Wailing Wall.

To get anywhere near it, you have to pass through a metal detector, and this one was operated by a young female soldier with long blond hair, red lipstick and a machine gun slung over her shoulder as casually as a handbag. She stubbed out her cigarette, gestured that I should raise my arms and began to frisk me briskly.

'Careful,' I muttered.

'Problem?' she snapped back.

'I'm pregnant,' I told her. 'You could be a little less rough.'

'Where your husband?' she asked rudely, glancing around.

'Why?'

'Not good to be alone here,' she told me.

'Really?'

'Perverts. Flashers. You be careful,' she said, patting my legs to check for explosives. She wasn't exactly selling Jerusalem as a tourist destination.

'Oh, okay. Thanks,' I told her. 'I'll be careful.'

At the Western Wall, I paused to watch a bar mitzvah – the coming-of-age ceremony for Jewish boys. A large family were giving whoops of encouragement to a skinny, bespectacled thirteen-year-old boy, who was reading earnestly from the Torah.

Around him, business went on as usual, with other Jews rocking backwards and forwards deep in prayer, and young soldiers, fresh into National service, prowling around. Taken away from their studies to safeguard their country for twelve months, they looked more intent on smoking and flirting. If there were troublemakers in the crowd I doubted they would notice.

With the warning of perverts and flashers still on my mind, when I arrived at the Jaffa Gate to walk the city's ancient stone ramparts to Lion's Gate, I attempted to tag on to the end of a German coach party (these were desperate measures for desperate times).

'This is a private tour,' a stout German woman told me.

'I didn't realize,' I lied. 'Does it really matter?'

'We paid for a private tour. Better you find another,' she barked.

I wanted to have company as I walked the ramparts that would take me past the Dome of the Rock, one of the most emblematic buildings in Islamic culture.

'I've heard that for female solo travellers it's best not to be here alone,' I told her.

She folded her arms and shook her head.

'And I'm not listening to the tour,' I added, which was true as it was in German. 'I won't bother you.'

'*Schleich dich!*' she hissed at me, which I later discovered meant 'Get lost!'

After that, it was tempting to hole up in my hotel room and if I hadn't been there on assignment I may have whiled away the remainder of my stay watching Israeli versions of *EastEnders*.[1] Instead, the next day I reminded myself that I had a job to do and hopped in a taxi to travel from central Jerusalem into Palestine on the West Bank, to visit Bethlehem.

'Are you sure this is the right way?' I asked, peering at a map in my guidebook. After yesterday, I'd decided that I should give maps another go. We'd veered off course, away from the main road, and were now travelling through the country lanes of a much quieter rural area. I wondered how long it would be before someone found my body in a ditch as it dawned on me that I hadn't told a soul where I was off to that day.

'Better this way,' the taxi driver told me, grinning at me in the rear-view mirror; his gold tooth glinting in the sunshine.

'But Bethlehem isn't in this direction, is it?' To be honest, I didn't really know. Rather it was a gesture, to let him know that I was on to him, so if he was indeed thinking of murdering me, he could think again.

---

[1] *EastEnders* is a British soap opera set in the East End of London, which has been broadcast on the BBC since 1985.

'No need to worry. We'll be there very soon,' he replied, switching on his radio as if he'd had enough of me. Questioning him further would be futile anyway. He was either going to kill me or not. Twenty minutes later, he dropped me exactly where I'd asked to go – on the corner of Manger Square – and wished me a good day like the nice man he was. As I got out of his taxi, I caught sight of a photograph pinned to the dashboard of him with his wife, two daughters and a son. I gave him a generous tip to make up for letting my imagination run riot.

Manger Square was more commercial than I was expecting, lined with market stalls selling olive wood worry beads and nativity scenes; I stopped to buy a Christmas tree decoration – a simple olive wood carving of Mary on a donkey. The focal point of the town is the Church of the Nativity where, as if by magic, raucous tour groups are reduced to quiet awe as they enter the underground chamber below the church where Mary is said to have given birth to Jesus. I'm not in the least bit religious, but with hormones kicking in and caught up in the romance of the moment I may have put my hand upon my growing bump in some kind of embarrassingly attention-seeking way as I watched pilgrims bend to kiss the spot where Jesus is believed to have been born. Really though, the place made little impression on me, and I was rather disappointed. Back outside, with senses restored, I tried to work out the direction of Milk Grotto Street.

'Can we help you?' I looked up to see two good-looking young guys in their mid-twenties grinning at me.

'Oh, I'm fine,' I told them.

'Are you lost? We can show you the way,' one insisted.

'I know where I'm going,' I said, heading off in what I hoped was the right direction.

'We'll come with you. Better than being alone,' the other said, walking in step beside me. 'Where are you from?'

My heart rate sped up. 'I'd really rather be left alone,' I said, taking a sudden change in course.

'Oh, come on. Be friendly,' the first one said, touching my arm. At this unexpected physical contact, I reeled back.

'Please leave me alone. I am with child,' I said, sounding as much like the Virgin Mary as I was able. I mean, come on. I was a lone pregnant young woman standing within spitting distance of the Church of the Nativity. Surely, even these two stalkers had standards and could see the irony of the situation? It worked like a dream.

'Oh, so sorry,' one said.

'Take care of yourself,' said the other.

I watched them slope off with their heads hung, like a couple of King Herod's henchmen.

Once I'd reached the Chapel of the Milk Grotto, I relaxed a little. Oddly, the entrance was at the front of the main chapel and visitors entered in full view of the congregation. In a moment of embarrassment at having been thrust into the limelight, I stuffed a large donation into the offering box, and then slipped quickly by the rows of women (and a few uneasy looking men), all fiddling with their rosaries and no doubt praying for good lactation.

Through a second humble door, I stepped into the smaller chapel where Mary is said to have breastfed Jesus and experienced the kind of warm glow you get from visiting a cosy country pub or your Granny's front room. I got a flashback to a grotto at Christmas where, aged five, I'd asked Father Christmas for a Tiny Tears doll. Was I having some kind of religious epiphany (doubtful)? Or had I been reduced to a bag of raging hormones (likely)? Either way, I was glad I'd visited.

For the next part of the trip, which would take me to Eilat and then into Jordan, I wouldn't be travelling alone. The partner of a colleague of Neil's (also on the team-building jolly) had arrived in Jerusalem and would be joining me. All I knew of him was that he was American, his name was Slim and that on my return from Bethlehem he'd left a friendly note at the hotel reception, which read, 'Shalom! Let's catch a cab to the airport together.'

At the designated time, I waited in the hotel foyer for my new travelling companion. When Slim arrived, I discovered that he was in his mid-fifties, wheelchair bound and a force of nature. Cracking jokes by the minute, I warmed to his live-life-to-the-full attitude and admired his tenacity. Being in a wheelchair was not going to hold him back. Going through security we stuck together, chatting all the while as he told me about his career as an art dealer in New York. I never once considered what an odd-looking couple we made, but others had.

'Can you tell me what your relationship is to this young woman?' a fierce-looking female security officer asked Slim.

'This girl? Well, I hardly know her. Picked her up at the hotel,' joked Slim, winking at me.

This was no time to be making jokes. Israeli security forces don't have time for them. Plus, Slim had intimated that I was a hooker. Not good. Not good at all. I was swiftly frogmarched into a nearby interrogation room by her. The last thing I saw was Slim mouthing 'Sorry' at me as the door clunked shut.

'What are you doing in Jerusalem?' she barked.

'Writing a feature. I'm a journalist.'

She narrowed her eyes. Foreign journalists were

generally considered nuisances in Israel at that time. I wondered if it might have been better to have said I was an escort.

'I write about travel,' I quickly added, but she'd already moved on.

'The man you are with, what is his name?'

'Slim,' I replied.

'This is not the name he has given.'

Of course, it wasn't. Slim was so obviously a nickname. 'Well, I expect Slim is what his friends call him. It was the name he told me, but it's probably something like Jim or James.'

From her expression, I could tell that it was neither.

'What is your relationship to him?'

'We don't have a relationship. He's the partner of a woman my husband works with.'

'And where is your husband?'

I hesitated, conscious of just how ridiculous my answer would sound. 'He's riding a camel in the Wadi Rum,' I told her.

'What for?'

I resisted the urge to say, 'F**k knows!' and instead explained that it was a team-building event.

At this information, she gave the slightest of smirks. 'And how long you stay in Israel?'

'I go to Jordan today. Could I please sit down?' I asked. I was starting to worry about missing the flight, which in turn was making me feel hot and a little faint.

'No,' she told me. 'Empty your bag.'

I did as I was told, putting my camera, guidebook, purse and few other bits and pieces on the table before her, which rather embarrassingly included something called a Lady-pee (a portable loo-bag for women on the move) that I'd only tried to use once (unsuccessfully) on a night bus in China.[2] Even so, it went with me, like a security blanket, on all of my travels. The guard picked it up and peered at it.

'What is this?' she asked, giving it a shake. The bottom of the pouch contained wee-soaking crystals, which, when jiggled, made a sound like either gravel or crystal meth granules, depending on your levels of suspicion.

From her belt, she took a knife, which was hanging in a sheaf next to a gun, next to a taser, next to some kind of blunt instrument like a small truncheon, and split the pouch open. The white granules spilled on the counter, which she took one sniff at, then swept aside.

'Pockets,' she barked at me.

Now I really did start to panic. I'd heard stories of strip searches, and even of internal examinations.

---

[2] I have made the name 'Lady-pee' up because it would be very embarrassing to be sued by a company that makes wee-bags.

I imagined the snapping sound that her rubber gloves would make as they were pulled on and thought I might cry, and then, in my pocket while searching for a tissue, I felt the crinkle of paper against my fingers. It was my twelve-week scan photo of Ernie that I'd recently taken out of my wallet to show Slim. I pulled it out and held the image out to her.

'Please can I sit down? I'm pregnant and you're frightening me. Here's a photo of my baby,' I said.

She took the black-and-white photo from me and squinted at it. In it, Neil and I had joked that Ernie looked like he was sticking his middle finger up at us, but she didn't seem to notice and I watched as her face softened to a smile. 'You can sit,' she told me.

I didn't miss the plane, although I was the last to board, and I could barely look at Slim, (who turned out to be called David), as he congratulated me about how 'no harm had been done' and that it would 'give me something to write about' (it hurts me to admit that he was right). No question though, it was Ernie who'd saved the day again.

From Eilat we travelled over the border into Jordan and on to Petra. I held it together for the two-and-a-half hours that we bounced along the imaginatively named Wadi Rum Desert Highway, a dusty road that cuts a line through a basin of sand-stone and granite mountains, known as the Valley of the Moon. Was this too bumpy a journey for little Ernie? Still reeling from the airport incident, I was tense and jittery, and I've never been more pleased to see Neil than when I clambered out of the jeep. So glad in fact that I burst into tears.

'I'm so stupid,' I wailed. 'What was I thinking of?'

'What's wrong?' he asked.

'It was too bumpy for Ernie,' I sniffled, putting a protective hand on the small but discernible new bulge to my stomach. 'What if I've hurt him?'

I blurted out a garbled version of my last forty-eight hours, which went something like this. 'Stupid German cow wouldn't let me join her tour group even though there were flashers. Taxi driver took me off the main road and I could have been murdered. Got followed by two men in Bethlehem and one touched me. Israeli security guards took me away. Thought I was a prostitute. Felt dizzy. Couldn't sit down. They cut open my Lady-pee with a knife.' This last sentence sounded way more painful than I'd meant it to and ended on a wail.

Understandably Neil's face portrayed a mixture of emotions, mostly alarm and confusion with the odd flicker of amusement. He asked if anything hurt (it didn't), if I was dehydrated (I wasn't), and then he insisted that I got straight into bed for a nap. He had one last unenviable afternoon of quad biking with his workmates to get though with a backside rubbed raw from riding a camel. I lay there for a while, and when calm, conceded that perhaps I may have got things a little out of proportion. There was a danger now that Neil might want to wrap me in cotton wool for the remainder of my pregnancy, which would really get on my nerves. Now safe and snug under a blanket, I accepted that I would have to rethink the way I lived my life from that day on. With my mum's concern ringing in my ears, for the first time I understood the overwhelming need a mother has to protect her child. My journey as a mum wasn't beginning in another six months, it had already begun and Ernie's wellbeing (even though he was the size of an apple right now) would be my number one priority from this day on. It was Ernie who had got me out of trouble twice in the last twenty-four hours. What a great team we were already, and I was determined to look after my new little travel buddy.

'Hungry?' I asked Ernie, stroking my growing bump.

I scanned the room-service menu. Should I have the traditional falafel served with an orange and pomegranate salad or the Western cheeseburger? I ordered both. Eating for two was surely the only excuse I needed.

2

Mallorca

A Lesson in Family Travel

Ernie, now named Josh, age three months[3]

10 November 2000,
Binibona, Mallorca

*Are babies at three months meant to be this active? The
practical baby bible says not although I'm thinking of chucking
it away as neither Josh or I seem to be in step with the weekly
'mother and baby' progress guide. Josh is way beyond (like a
baby on speed) and I'm way behind – not yet puréeing butternut
squash and freezing it in ice cube trays so that I am 'WELL
PREPARED' for a zombie apocalypse. Napping at intervals of
twenty minutes (if I'm lucky), for the rest of the time Josh is on
the go, preferring not to sit but to stand as I support his weight,
so that he can see what's going on. Lively though he may be, our*

---

[3] Ernie was named after Ernie from Sesame Street. If we'd known we
would go on to have two more sons we might have stuck with the
name – adding a Bert and possibly an Elmo to our family.

first flight went pretty smoothly. Josh didn't cry once but shouted at the cabin crew every time one passed us by, emitting a friendly 'wah' to catch their attention. Lunch was served late as a result, as not one of them could resist a quick cuddle. Sticky moment for Neil, who bravely said he'd have a go at changing Josh's nappy on the pull-down changing station in the loo, when he managed to get the changing bag's strap (which he had slung around his neck to carry) trapped in the hinge as he pulled the station down. He now has a lovely rope burn on his neck that looks like he has tried to hang himself. Josh survived unscathed.

Another slight problem on arrival as our one and only check-in bag had gone astray. Luckily though we'd bought two bottles of duty-free champagne, which came with a free teddy bear, so arrived at the hotel clutching only these and Josh, like a couple of rock star parents.

Once at the hotel, the sight of a) luxurious toiletries in bathroom, b) chocolate on my pillow and c) a room-service menu brought tears to my eyes. And Neil got very excited when he discovered that behind the gorgeous silk curtains there is an electric black-out blind that turns our room as dark as a coal pit. A travel cot has been set up next to our bed that is decorated with clown faces, which would scare the crap out of me, but as Josh has not yet read the book It, *we're hoping all should be okay.*[4]

---

[4] It *is a 1986 horror novel by Stephen King, in which Pennywise, an evil clown, preys on young children.*

MALLORCA

nappies	bottle x 2
travel changing mat	teething ring
wet wipes	teething gel
nappy cream	sterilizing tablets
spare clothes (Josh) x 2	blanket
spare top (me)	dummy
breast pads	spare dummy
emergency formula	emergency dummy x 3

Since becoming a mum, I'd started making lists, terrified of leaving home without essential items. Although on one occasion, I must admit to having gone a few strides down the road without Josh! I was sitting on the floor, with Josh on his *Sesame Street* play mat happily punching the living daylights out of the Elmo soft toy that was dangling above his head. Next to us was the mountain of stuff that I was about to try and stuff into my hand luggage and I was already nostalgic for the days when a passport, lipstick and book (and a Lady-pee of course) were all that I required. I hadn't really considered what travelling with kids in tow would be like. I had been much like any other child-free traveller in that my heart sank when families sat next to me on planes, anticipating just how noisy their kids might be, particularly if they were babies or toddlers. I wrote mainly about off-the-beaten-track adventure travel, so rarely encountered families. Although in Pushkar, India, while on a camel safari in the desert, I had met one Dutch couple with three young ginger-haired kids between the ages of five and ten years. They were slathered head to foot in what looked like whale fat, and I thought to myself, 'Oh, for God's sake, take them to Centre Parcs.' Did I see myself with a child of my own travelling the world? Not really.

And now here I was, just three months after Josh's birth, packing for my first trip with my new baby boy. The visions I'd had of myself as a cool travelling mummy, wearing something effortlessly boho and sauntering to a departure gate with Josh gurgling happily, swaddled to my calm beating heart in an ethnic wrap ('Oh, just a little something that I picked up on my travels,' I would say to other admiring mums) were quickly ebbing away. Number one, Josh, from the word go, hated any kind of bodily restriction. We'd already tried four different types of baby carrier, causing Neil to exclaim, 'Well, he'll just have to walk.' Number two, none of my cool clothes fitted me, mainly due to the fact that my breasts were enormous, in that horrible lumpy way caused by breastfeeding that no one ever warns you about. And now I had all this kit to lug about, in a bulky black changing bag that looked more like something an electrician might carry (it came with the pram and for economy sake I felt that I must use it).

All that said, Josh was a happy baby with an easy disposition that had made the first few months of new parenthood an absolute joy and because of this I wasn't particularly worried about taking him on his first plane journey. Plus (and this was a massive plus) I was so looking forward to having meals cooked for me, time to read and swim, and having Neil around. His job at that time required him to be out at 7a.m. and, if there was a fair wind, home at 8p.m. When he was around, however, he was one hundred per cent present; giving me and Josh his full attention and work was always left where it belonged, at the office.

I'd been invited back to stay at a historic sixteenth-century *finca* (farmhouse) in the north-west of Mallorca that I'd written about previously. The owner, Pete, had been thrilled with the coverage, which had begun with just one commission and had ended up as six, including a huge spread in the airline magazine of (the now defunct) *GB Airways* and a feature in *The Sunday Telegraph*, both of which had brought him lots of business. The place was beautifully restored, very luxurious and also adult-only. In these features I'd written that 'Only the tinkle of a distant sheep bell interrupts the tranquillity of the location' and that it was a place to 'curl up like a sixteenth-century farm cat, to enjoy the most blissful of undisturbed sleep'. I'd once asked Pete if he'd ever considered having families to stay and his reply had been simply to wrinkle his nose (no doubt imagining a used nappy shoved into the artisan rattan baskets that served as bathroom bins).

'I'll make an exception,' he told me on the phone. 'You can bring the baby.'

The fact that Pete was persistent in referring to Josh as 'the baby' did worry me a little, but then I thought back to the beautiful suites with luxurious canopied beds; the views out to almond and olive groves, and beyond to the Serra de Tramuntana; the free-standing bath big enough to float in; the delicious breakfasts we'd enjoyed on the *finca*'s sunny terrace amid the fig and mulberry trees. Pete had told me that Mallorca was enjoying an Indian summer, with temperatures by day still hitting 23°C, and I was seduced. After all, Josh wasn't a howler and so surely it wouldn't matter if we were the only guests with a child?

'Well, if you're sure,' I said. 'Thank you. We'd love to come.'

We arrived early evening. 'I'll be serving drinks on the terrace in thirty minutes,' Pete told us. 'And you'll join us for dinner at 9 o'clock, won't you?'

'Oh yes. That would be wonderful,' I said, giddy at the thought of having a social life once more. Dinner at 9 o'clock, like sophisticated people. That was the time when we'd normally be going to bed.

'I'm just going to have a soak,' I trilled to Neil, as I swanned into the enormous bathroom with its free-standing tub. 'Oooh and open one of those bottles of champagne. I'll have a glass in the bath.'

'Hmmm. One problem,' Neil said.

'Problem?'

Neil lifted Josh up and dangled him in the air.

'Oh.'

Crazy as this may sound, just for a couple of minutes I'd forgotten all about Josh.

'Perhaps you could feed him early?' suggested Neil.

'What?' I screeched. 'Are you out of your mind?'

Back then, in those early months of motherhood I thought that my only chance of survival was routine. Now, three children later, it feels like looking through the window at a mad woman. 'It would be a catastrophe!' I continued. 'Then his next feed would be out. Then the next one. And what about my boobs? I'd either have not enough milk or too little. Do you want to be up at 3a.m. trying to help me express milk, well do you, do you?'

Neil gave me a look that expressed the full horror of that particular scenario. Plus, the stone walls to our suite, once a farm worker's cottage, were over a foot thick, which the baby monitor had no chance of penetrating. After making my apologies to Pete, we ate dinner in our room and went to bed at 9p.m. A social life would have to wait.

The next day, however, had a new rosy glow about it. After a feed at 11p.m. Josh had slept until 7a.m. (actually 6a.m. GMT, but neither of us mentioned that, not wanting to burst the smug bubble).

During breakfast, I watched Pete do his rounds to guests. Was it my imagination or was one guest gesturing over to us and enquiring about the legality of a baby on the premises?

'Stop being paranoid and relax,' Neil told me, when I whispered what I'd seen.

Soon it was our turn to be schmoozed. 'I trust all is well,' Pete boomed. 'And that you have all you need for "the baby"?'

'The baby is fine,' said Neil, and winked at me.

'Then, I'm sure you won't mind if I borrow your good wife for half an hour. I have some new features around the place I'd like Kate to see, should the opportunity arise for more coverage.'

Neil raised his eyes at me. We'd both known that I wouldn't get away with visiting without at least one sales pitch, and as the man had flown us here and we were staying as his guests, it felt like the least I could do. Josh had just finished breakfast, which had been given discreetly under the cover of a napkin, so time was my own. 'I'll come now,' I told him.

Following Pete to view the latest of the original farm dwellings to be refurbished, I listened to him talk about the new autumn menu he would be launching the following week and when he asked my advice on an idea he'd had for a wine-tasting and cookery weekend, I gave myself a mental pat on the back. Look at me, I thought, still living my travel writing life. It wasn't until I caught sight of myself in the antique-mirrored wardrobe of the new suite that I was reminded of my real purpose: a producer of milky drinks. I had forgotten to clip my feeding bra back up on my right side and my breast was wobbling around like an underset milk pudding. My cotton shirt, caught in the light, was a little more transparent that I'd realized, but that wasn't the worst of it. My breast pad, which had been carefully placed on my lap while I Josh fed, was stuck on the crotch of my shorts like some kind of weird codpiece. It made a noise like Velcro as I peeled it off. I crossed my arms over my chest and hoped Pete hadn't noticed, then went back to my room to nurse my embarrassment and have a self-indulgent cry.

'Never mind,' said Neil, when I told him. 'No different to going topless by the pool, really.'

'Except that I was in a business meeting of course.'

'In a bedroom.'

'Could it be much worse?'

'Both tits out?' Neil suggested.

In the swimming pool, Neil was singing, 'Three little ducks went swimming that day, over the hills and far away. Mummy duck said quack, quack, quack, quack, but only two little ducks came back' on repeat, and Josh was lapping it up, hitting the water with his fists and squealing with pleasure (even though, on reflection, it is a rather sinister song). I watched as a nearby guest frantically searched for earphones in her canvas tote and told myself that I should take Neil's advice to stop worrying about what other people might think about us being there. Reaching for the camera, I zoomed in on Josh's happy face. But wait, what was that bobbing about just behind him that looked a little like a fun-size mars bar, but was the colour of a Caramac? Oh, shit. Literally. I searched for a nappy bag and when I was sure no one was looking shoved it down my bikini bottoms and casually entered the pool.

'Thought you were going to read?' said Neil. Our agreement was that we would work shifts, and every thirty-minutes or so we'd swap until Josh was tired enough to fall asleep and then wahey it would be snooze o'clock for us all.

'Divert attention,' I ordered.

'What?'

'Poo. In. The. Pool.'

'Oh Jesus. You sure it was Josh? He's got a swiming nappy on?'

I glanced up at the woman by the pool who was reading *The Sunday Times*. 'I'd say it's likely, don't you?' I replied. Fast forward three months to Centre Parcs, where we watched as the entire pool was evacuated so that a man, wearing the kind of equipment you'd associate with deep sea diving, could catch a floater in a net. I'd say that, concept wise, the swimming nappy was a genius invention. Design wise, it could do with a little more work.

A Navy SEAL could not have completed operation poo-in-the-pool more professionally. The bag was whipped from my briefs as I swam, and I timed it perfectly on the outwards breaststroke to scoop up the offending article. Then, while pretending to enjoy the sun on my face, I paused, and underwater, as deft as a balloon modeller, tied a knot in the bag and tucked it back in the front of my bikini bottoms. Neil returned to the sunbed ten minutes later and glanced hopefully at his book. 'You must be bloody joking,' I told him.

Of course, this catalogue of new parent disasters is only one side of this story. With Josh dozing in his buggy we strolled to the local town, where in the square we drank sangria and ate platters of cold meats and local cheeses under a trellis dripping with bougainvillaea, while a steady stream of old ladies in black came to coo over Josh, who took such a liking to one with dyed red hair and green eyeshadow that he cried when she handed him back. These were wonderful days spent gazing at our firstborn, marvelling at every little movement, noise and smile he gave us, totally absorbed and smitten with our son (but a chapter on that kind of new parent euphoria would have been mind-numbingly boring for anyone other than us). Having to secrete a poo in my bikini bottoms was a small price to pay for time spent as a family – the first of so many wonderful adventures, with not just one but soon three lovely sons.

3

Thailand

A Lesson in Recovering from Disappointment

Josh age three and Ben two

New Year's Day, 2004,
Choeng Mon Beach, Koh Samui

I'm still giggling about last night's New Year's Eve adventures. After an idyllic day on the beach (Josh is now on first-name terms with the ice-cream seller, the man who runs the beach café and the girl who braids hair), we took a dip at dusk in the hotel's boat-shaped swimming pool and watched the bats swoop in to swirl around the coconut palms. The boys were giddy at the thought of being allowed to stay up until midnight, daring each other to swim under a cascade of water, which flows from the trunk of a stone elephant fountain, and singing 'Na na na na .... Batman' each time a bat flew into view.

I'm so glad we opted out of the hotel's 'Gala' buffet and instead went barefoot in the sand at the no-frills restaurant at the end of the beach. Our table had a view out to the tiny offshore island that the boys have nicknamed 'pirate island', which we kayaked to this morning. After a dinner of coconut prawns and chicken skewers, with much excitement the owner announced that the show would soon begin. We expected the usual Thai dancers, perhaps a few musicians thrown in to drum in the New Year. What we didn't expect was a ladyboy show. Josh surprised me by cottoning on pretty quickly that perhaps all was not quite how it seemed, while Ben remained oblivious.

'Mummy, is that a man or a lady?'

Explaining that ladyboys may either be cross-dressers or transgender was far too complicated a conversation, so I settled on, 'Well, Josh. It's a little tricky to explain, but I think at one time she was a man but is now a lady.'

'Like a transformer?'

'Sort of.'

During the interval, Josh and Ben chased each other around on the sand – high on mocktails called Coral Reefs and the thrill of being up so late – and the ladyboys in their sequined dresses and feather boas came over to pinch their cheeks and coo over them. 'My name's Benny,' we heard Ben say. 'Oh, Benny! You so cute!' came the reply.

'Where you from, Benny?' another asked.

'Windsor.'

'Where that, honey?'

'England,' said Josh, to help his little brother out. 'We live next door to a queen.'

Predictably his answer caused the gorgeous ladyboys to squeal with laughter.

It was a lovely moment, full of innocence, and one we'll tease them about for years.

~

This was our first long-haul trip with two young children in tow, and I couldn't wait to be striking out beyond Europe again and waking my senses up in the temples and street markets of Thailand. Although Neil and I had backpacked around parts of South-East Asia some years ago, Thailand would be a new destination for all of us and I was never happier than when my bags were packed and an adventure was before us. As our departure date grew nearer, Josh and Ben's excitement reached fever pitch, too, which I assumed was my eager anticipation rubbing off on them. I wasn't particularly fazed by the eleven hour flight before us because I had it all worked out: a visit to a play park on the afternoon of our departure to wear them out; a carefully chosen flight time at 9p.m. so the boys would be tired but not overwrought; pyjamas to change into; their favourite blankets and teddies to snuggle up with; and cartoons to watch on the inflight entertainment if all else failed. I was imagining the film I'd watch and the glass of wine

I'd enjoy as soon as they were sound asleep. So, it came as something of a surprise when Ben was not compliant. The minute we boarded the plane he began to cry, and I could sense the immediate unease of my fellow passengers as they shifted in their seats and rolled their eyes.

'Want to go home,' Ben screamed, going stiff as a board as I tried to do up his seatbelt.

'Not an option,' I told him, adhering to the belief that it was always better to be honest.

'Whaaaaaaaaaaaaaaaa!!!' screamed Ben.

'Got a screamer, row thirty, seat C,' I heard one of the cabin crew helpfully mutter to her colleague as they passed by. I tried once more to fold a rigid Ben in two. 'Go 'way, Mummy,' he wailed.

The woman behind us requested to move seat. 'I haven't the patience for crying children on a flight of this duration,' I heard her say. My stress levels were soaring to ten thousand feet and we hadn't even left the runway. I looked to Neil for help, but he'd drawn the long straw and was busy helping a compliant Josh into his pyjamas. So, I decided to take control. Standing up, I addressed my fellow passengers. 'Can I have your attention, everyone?' I began.

'I know you're all thinking that my son is a nightmare and he'll be screaming for eleven hours, but he won't. So, if you could all stop tutting and complaining while I get him settled that would be really helpful. Okay?' Shame-faced, most nodded. A young couple, who were most likely setting off on some wonderful backpacking adventure, gave me the thumbs up. Neil paid me no attention, which I put down not to embarrassment but rather an acceptance of my rather feisty way of going about things, but I later discovered he'd already inserted earplugs. While we were taxiing down the runway for take-off Ben fell asleep and didn't wake up until we landed in Bangkok.

Josh was keener to stay alert, but after watching a film, he too nodded off, waking an hour before landing. As long-haul flights with young kids go, while sleeping upright in economy, this had been a success and I was feeling just the tiniest bit smug as we tucked into breakfast.

'We nearly there?' Josh asked.

'Yep. Are you excited?' I asked.

'Will Noddy be here to meet us?' he asked, bouncing in his seat.[5]

'Noddy? What do you mean, Josh?'

'Silly Mummy,' he said chuckling, and lifted up his Noddy soft toy as if to explain.

'Yeah, I know who Noddy is, but why do you think

---

[5] Noddy is a fictional elf-like boy, created by English author Enid Blyton between 1949 and 1963, and he lives in a place called Toyland. Several animated television series, based on Blyton's Noddy books, have been made.

he'll be here to meet us?'

Again, Josh cracked up. And then suddenly the reason for their pre-departure excitement fell into place. The constant recent request for *Noddy's Adventures in Toyland* at bedtime should have been a clue. I looked at Josh – with his fuzzy bed hair, flushed cheeks and shining eyes, who was literally shaking with anticipation – and panicked. 'Probably not at the airport,' I said, before turning to Neil to fill him in on the fact that the boys thought we'd brought them to Toyland, not Thailand. Easy mistake to make, given that a few months earlier we'd been to Euro Disney (as Disneyland Paris was then known), where they had met Mickey Mouse, Donald Duck and Pinocchio. In my three-year old's fertile imagination, a trip to Toyland to meet Noddy was certainly a possibility.

'Oh, shit,' Neil said helpfully.

'Just go with it for now,' I said. 'We'll let him down gently after we land'.

This wasn't my usual way of going about things but I thought it likely that Josh would forget about Toyland once he clapped eyes on Bangkok in all its chaotic, colourful splendour.

As we were getting ready to disembark, the woman who had requested a change of seat at the beginning of the flight grabbed my arm. 'Well done,' she said, nodding towards my *enfant terrible*. 'We didn't believe you when you said he'd stop crying, but he's been good as gold.' I allowed myself just one more brief moment of smugness, before Josh tapped her on the arm. Ever the socialite, he was keen to join in with the conversation.

'Tubby Bear lives here,' he told her. 'And Big Ears, who is Noddy's best friend.'

'Oh,' she said. 'I didn't know that.'

Josh looked amazed. 'I hope you don't see that naughty goblin,' he told her. 'But watch out because he lives in Toyland too.'

She raised her eyebrows at me.

'Long story,' I whispered.

After navigating passport control, customs and baggage reclaim (where I may have been heard to shout 'Get off the belt' more than once) we headed towards the taxi rank.

'Noddy is a taxi driver,' Josh told me, scanning the queue of waiting cars. 'Is he here?' I decided to pretend I hadn't heard. 'Remember when the naughty goblin painted his yellow taxi?' he persisted. 'PC Plod put him in jail and . . .'

A purple taxi painted with tropical fish advertising Koh Samui, where we would be travelling to in just a few days, caught my eye. 'Get that one,' I instructed Neil.

'Oh wow, look at this car. Much better than Noddy's, right?' I said as we clambered in.

En route to our hotel, Josh cottoned on that something was wrong. 'Doesn't look like Toyland,' he said, peering out of the window at the gridlocked traffic. He was right. There were no mushroom-shaped houses, no giant flowers, no cars with smiley faces and there was no way that I was going to get away with this for a minute longer.

'Oh, for God's sake, Josh. We're not in Toyland!'

'What happened to letting him down gently?' cried Neil, as we watched Josh's lip begin to wobble.

Truth was, I was too tired for this nonsense, and already sick of hearing about Noddy.

'Better he knows,' I snapped. 'Josh, we're in Thailand, not Toyland and it'll be much more exciting. Okay?'

'But you said . . . '

'I never once told you we were coming to Toyland. You just misheard.'

I could see that Josh was trying to be brave. I'd seen this kind of disappointment before – at Christmas when he didn't get picked to be a king in the nursery-school nativity; on discovering that he was too short to go on the Dragon Rollercoaster at Legoland (even after Neil had stuffed paper napkins into his trainers in an attempt to add an inch) – and it always started with silent weeping. I watched the tears plop onto his cheeks. 'Benny, we're not in Toyland,' he whispered to Ben, choking on a sob. Ben didn't seem bothered. He was quietly assimilating his surroundings and deciding if he liked them (we'd soon know about it if he didn't): watching the coloured lights that decorated our taxi driver's windscreen flash from orange to purple to green, and the smiling tin cat that nodded its head on the dashboard and, through the window, the neon billboards advertising the likes of Pei Wei river noodles and Thai massage, all interspersed with pictures of the King.

'I'll make it up to you,' I said, giving Josh a squeeze. He shook his head, folded his arms and turned away from me.

Okay, so looking back, Bangkok with its traffic congestion, terrible pollution and infamous red-light district was perhaps a curious choice for the start of our first long-haul trip with two such young children. In a cloud of carbon-monoxide, with horns blaring all around us, and Josh quietly sobbing next to me, I stared out of the window at a billboard and tried to decipher what the company 'Sodick' might sell. Suddenly Toyland looked pretty attractive.

A few hours later, on the terrace of our hotel with a view to the Chao Phraya river, I wearily watched Ben trying to eat ice cream with chopsticks. He'd decided by now that Thailand was fun. Unlike Josh, who was stubbornly in mourning for Toyland.

'Why don't you have a go, Josh,' I suggested, handing him a pair of chopsticks.

'It's stupid,' he said, glowering at the sticks. 'And Benny's made a mess.'

That was an understatement. By now Ben was wearing more than he'd eaten. 'But fun!' I said, attempting to scoop some from Josh's bowl with my own pair.

Josh shook his head. Sleep, was what we all needed. Things would look different in the morning, and I was hoping that all thoughts of Toyland would vanish once we set off out to explore.

Next morning, though, Josh remained reluctant to let me off the hook, and there was a sticky moment upon leaving the hotel when I had to wrestle Noddy from him on the pretext that he might get lost (I wished) and would be safer staying behind to look after our luggage. It seemed to me this soft toy was driving a rift between me and my firstborn.

'But Noddy will be lonely. I'm his only friend,' Josh told me.

'What's that, Noddy?' I asked, holding Noddy up to whisper in my ear. 'You've got jet lag? Oh, no. Best if you have another sleep, then. Don't you think, Josh?'

Reluctantly Josh helped me tuck Noddy back into bed, and we headed out to catch the sky-train, which speeds in airconditioned pollution-free bliss above Bangkok's gridlocked streets. If anything could take his mind off Toyland, I hoped that it would be the

Chatuchak Weekend Market – a manic sprawl of around seven thousand stalls where you can buy anything from antiques to animals. Keen to part with their holiday loot, the boys were soon in pursuit of flashing Spiderman sandals, Scooby Doo t-shirts and, in Ben's case, chipmunks. He peered into their cage and, like a well-drilled circus act, they took their moment to perform, turning somersaults and chasing their tails to illustrate just how much fun they'd be if he took them home. Ben looked at me hopefully. 'Perhaps another time,' I told him.

Josh had his eye on a fake Chelsea football kit, so I handed him the equivalent of three pounds in baht. 'All I got,' he said waving the notes at the stallholder.

'Businessman,' the man commented, high-fiving Josh, who by now was cheering up a little.

I was anxious to soak up as much of the city as possible in the two days we had there, and so flagged down a tuk-tuk to take us to Wat Pho, one of Bangkok's biggest temples and home to Thailand's longest reclining Buddha that stretches some 46 metres and is covered in gold leaf. The most impressive part of this massive shrine are the feet.

'Want to tickle the toes?' I asked Josh.

'Me!' said Ben, holding his arms out for Neil to lift him up to have a go.

'Tickle, tickle,' I heard him chuckle, while Josh remained unimpressed.

'Do you think Noddy is awake now? Shall we go back to the hotel and get him?' he asked.

'Over my dead body.'

'Say again Mummy?'

'I said better not to disturb Noddy.'

We needed another distraction and fast. 'How about we set sail on our very own pirate ship?' I suggested, surely the perfect adventure for two wannabe buccaneers. Luckily, in Bangkok, long-tail boats are easy to come by, as common as black taxis in London.

'Ready to set sail, me hearties?' I asked the boys, once we'd flagged one down from the riverbank. By now, Josh and Ben's cheeks were so flushed from the 33°C heat I thought at any minute they might combust, leaving only two pairs of smouldering sandals.

'Want to be captain?' the captain asked Josh, picking him up and placing him on his lap to steer. Josh beamed back at us. I didn't need to worry that Ben might be jealous. There was no way on earth he'd ever want to sit on a stranger's lap. He was much happier cosied up with us to watch life on the river float by. We sailed away from the busiest stretch of the Chao Phraya river and into the Thonburi canals lined with houses built on stilts, where kids not much older than Josh and Ben were doing star jumps into the water. Their grandparents, on child-care duties, kept one eye on their charges while tending to river gardens on small pontoons, and waved to us as we passed by. On the surface Bangkok appeared brash, chaotic, grimy and perhaps even a little hostile, but we'd only been in the city for less than a day and I was already discovering small pockets of tranquillity such as this, and learning that the people were gentle and friendly, always ready with a smile, particularly for our sons.

At the Royal Barge Museum we jumped ship to see the world's largest dugout boat, the 50 metre-long Suphannahong (golden swan), which needs a crew of fifty oarsmen to propel it through the water. From the brow, the Hongsa, a mythical swan-like bird reared up. It was beyond impressive, but something else had caught Josh's eye.

'Who's that?' asked Josh, pointing up to the figure-head on another large barge. It was Garuda, a mythical bird-like beast found in both Hindu and Buddhist religions, and the emblem of Thailand.

'Oh, that's Garuda, otherwise known as eagle-man,' I replied casually, with a plan quickly forming.

'Does he have super-powers?'

'Oh yeah. Loads,' I said, realizing that – pardon the pun – I'd have to wing it a little. 'He's king of the birds and can fly, and he's super-strong like a warrior. See how he's wrestling that snake? And he sees everything with his eagle eyes.'

Josh couldn't take his eagle eyes off him. 'Does he have friends?'

'What like other superhero friends you mean?'

Josh nodded.

I quickly thumbed through my guidebook, and settled on Ganesh, the elephant God, one of the best known and most worshipped deities in Hinduism. 'Well, yes, there's Ganesh. He's part elephant, part man.'

Josh's eyes grew wide. 'What can he do?'

I wasn't sure but resisted the urge to say spray water and trumpet. 'Well, he is really clever . . .'.

Thankfully Josh interrupted. 'Who's that?' he asked, pulling me further along the pontoon.

He'd spotted Hanuman, the Hindu white monkey God, looking fierce with teeth bared, as he soared from the front of another vessel. This was the cheeriest I'd seen Josh since we'd arrived. 'That's another friend of Garuda's,' I told him.

'I like Garuda,' Josh sighed. 'I could be Garuda, and Benny could be one of his friends.'

Back on the long-tail boat and heading back to the hotel, Ben spotted a floating shop piled high with goodies, from star fruits, green coconuts and mangos, to sacks of rice and souvenirs. 'Thirsty,' he told me, although I knew this was really an excuse to take a closer look. Our captain gestured to the conical-hatted boatkeeper, who oared her shop over so that we could browse her wares. Ben immediately homed in on a pile of small carved wooden replicas of the lady's floating boat shop. 'I like that one,' he told her. She held it up and laughed, pointing to herself and then to the tiny wooden lady on the toy boat. 'Same, same,' she told him.

Meanwhile, Josh had scrambled along the deck and was jumping up and down. 'There's Garuda!' he shouted, pointing to a bag of plastic figures, nestling amid the pots and pans in what looked like the household department.

'I doubt it,' I said, but gestured to the boatkeeper that I would like to take a closer look.

Within the bag, there were six plastic superhero-type figures with terrifying faces in a mix of garish colours, one of which had wings and would pass for Garuda if you squinted.

'Do you know, Josh? I was wrong. This is Garuda,' I said, and Josh clapped his hands together.

'Three for me, three for you, Benny!' he announced, beaming expectantly at me.

I handed over the Baht equivalent of two pounds.

'Noddy who?' chuckled Neil, as we watched Josh hold Garuda aloft. 'Gaaaaaaaa,' he shouted. 'I am eagle-man. Come on, Benny!'

Ben was less keen to join in. He was still examining his boat. The idea that he had in his hand a tiny version of the boat that he'd just bought his boat from, complete with the boatkeeper in her conical hat, was still clearly blowing his mind.

The holiday went better from that moment on. Back at the hotel we swam in the pool and ordered ice creams that Ben again insisted on eating with chopsticks, and this time Josh joined in. Early that evening, we wandered through the night market, making sure to steer clear of the girly bars and seedier side of the city that screamed for attention just a block away. The boys bought souvenirs – recycled Chang beer cans cleverly fashioned into tuk-tuks and small wooden frogs, which made a croaking sound when you stroked them along their backs with a stick. I noted, with some relief, that Noddy had not been invited on this excursion. At bedtime, I couldn't resist a small gloat. 'You were a good boy today, Josh. I know you were disappointed because we weren't in Toyland, but because you cheered up look what happened. You found Eagle-man Garuda and the others. How great is that?'

Garuda was tucked up in bed with Josh, and he glanced at his new best friend.

'Noddy can't fly and he isn't good at fighting,' he whispered, looking over to where Noddy had been casually flung on the floor.

'That's true.'

'And his friends don't have special powers.'

'No, they don't.' I was pretty sure Big Ears had the ability to cast magic spells, but I kept that to myself.

From then on Garuda was Josh's constant companion on our travels through Thailand, and poor Noddy rarely saw the light of day. In Koh Samui, Garuda snorkelled in Ang Thong Marine National Park, rode elephants through the jungle in Chiang Mai and ate barbecued shrimp in the night market in Bangkok. And on New Year's Eve, Garuda even enjoyed a ladyboy show.

4

Cambodia
A Lesson in Sensitivity

Josh age four and Ben three

8 *April 2005,*
Phnom Penh, Cambodia

Without a doubt the highlight of the day for the boys was the unexpected opportunity to cry, 'Yuk! Stinky poo!' at the top of their voices, when an elephant deposited several large dung balls the size of bowling balls in front of the bicycle rickshaws we were riding in. As our drivers expertly zig-zagged their way through this rather smelly obstacle course, Neil took photos of Ben and me in our rickshaw, holding our noses and pointing to the elephant's bottom. That's a picture destined for a frame.

From the safety of our moving vehicle and snuggled up to me, Ben is Mr Sociable, waving at everyone, from the woman dishing out noodles from her kerb-side stall to the teenagers cruising on their motorbikes and everyone shouts back, 'Susadei!' (Hello!). In heat pushing 40°C, the boy's chubby tomato-red cheeks are causing quite a stir everywhere we go, and I've yet to see another Western family travelling with such young kids.

When we wandered through the market on a food tour, the stall holders all reached out to touch the boys gently on their arms. 'For good luck,' our guide explained. Ben was expert in dodging these advances and treated it like a game of tag, while Josh embraced the experience, beaming widely at everyone he passed and pausing to admire their produce. One stallholder even presented him with a mango, which he passed to his private secretary (me).

As we were going out to dinner tonight, Josh asked if we were going to a temple, as if that's all we ever do (to be fair, though, they've visited three today). Instead, we took them to drink mocktails out of elephant-shaped cups amid the faded grandeur of Raffles Hotel Le Royal, where I caught Ben sneaking peanuts into the pocket of his shorts. 'For the elephants,' he whispered.

~~~

At four, Josh was like any other kid of that age, in that he liked to stare at anything or anybody that was even the tiniest bit different. Not the sneaky glances that older children learn to take, but that awful full-on gawping, with mouth open and often accompanied by pointing, that made my toes curl. I thought I'd made it clear to Josh that everyone had the same depth of feeling and that he could upset them by gawking so much. Then, just before our departure to Cambodia I had an embarrassing moment in Tesco when Josh asked loudly at the checkout, 'What's that stinky smell?' Said smell was emanating from the old bloke standing in front of us in the queue who, it must be said, looked and smelled like he hadn't acquainted himself with a bar of soap for some time. The man turned and glared at Josh, who, I could tell, from his rapidly colouring cheeks, was regretting his question. Josh wasn't an unkind boy. He was a kind boy who hadn't yet learned to disguise his curiosity. It was obvious that I needed to address his uncensored observations before our trip to Cambodia where accidents from landmines (first laid during the ousting of the Khmer Rouge in 1979 and then used widely by all sides in a war that lasted until 1998), had resulted in an estimated forty thousand people losing limbs, making Phnom Penh the amputee capital of the world. The conversation went some-thing along these lines:

'Josh, so I need to talk to you about some of the people you'll see and meet in Cambodia.'

'Why? Are they different?'

'No, just like you and me, but during a really bad war bombs called landmines were put on the ground and never picked up.'

'Are they still there?'

'In some places in the countryside, but we won't be going anywhere near them.'

The truth was that although Cambodia had made great headway in recent years in clearing these mines, there were thought to be millions of these death traps still littering the countryside.[6] Not wishing to alarm Josh, I kept this fact to myself.

'Anyway, some poor people didn't know they were there and stepped on them,' I continued.

'What happened?'

'The mines exploded, and they lost a foot or a leg, sometimes an arm, and sometimes they were killed, and when we're in Cambodia we'll see people who have lost an arm or a leg and you're not to stare or point, okay?'

Josh was quiet but nodded.

'Because that might upset them. Remember the man in Tesco? He wasn't happy was he?'

'The smelly one?'

I rolled my eyes but nodded.

'He was cross,' remembered Josh correctly.

'Exactly Josh. Just try to think a little more about other people's feelings, okay? Any questions?'

Josh shook his head, and I congratulated myself on a tricky subject covered.

---

[6] The Cambodian Government has since made a pledge to be landmine free by 2025.

Arriving in Phnom Penh I couldn't have been more excited. Cambodia had been on my wish list for years. I wanted to see the temples of Angkor Wat before the country was overrun with tourists, and in 2005 it was still a good few years off the boom that has made it one of the most visited countries in South-East Asia. We jumped in a taxi for the short ride from the airport to the centre of the capital and Josh and Ben were glued to the windows, pointing out all they could see – bicycle rickshaws and tuk-tuks; street hawkers selling flowers, baguettes and fruit, piled into perfect pyramids; a woman carrying baskets of ducklings; stray dogs dodging traffic, and, and . . .

'Oh wow!' shouted Josh.

'What is it?' Neil asked.

'I've just seen a man with no head riding a bicycle.'

'You haven't.'

'I have. He was there. And he had NO HEAD.'

Josh's eyes looked ready to pop right out of his head, and Ben was scrambling over him hoping to see the bicycle-riding man without a head, too.

'Josh, there is no way,' I began.

'But you said we'd see people who had lost arms and legs . . .' he interrupted me, looking cross.

'Yes, but not heads!' I said. 'Think about it, Josh. How would he breathe?'

54

While Josh contemplated this, I realized in retrospect that my little pre-departure talk had probably been pretty alarming for him, causing his imagination to run to all sorts of absurd places. That's why his nose had been pressed up to the glass on the taxi journey. I gave him a cuddle. 'Easy mistake,' I whispered, worrying over how much stress I'd caused my four-year-old. 'But, can I just say, that there is no way that you'll ever see anyone without a head riding a bicycle.'

As it was likely that Josh would encounter someone with a missing limb soon enough, now seemed as good a time as any for a very gentle reminder to be courteous. 'By the way, Josh. If you had seen a man with no head, which you didn't, it would have been better not to shout about it,' I said.

'Although, having no ears he wouldn't actually have heard,' pointed out Neil.

As a surprise for the boys, I'd arranged a tour of the city by bicycle rickshaw for the morning of our first day. I love rickshaws – the gentle pace the bicycle affords, the proximity to all that is going on around you – and I'd often used them as way of zipping around cities and towns on previous child-free trips to South-East Asia. The plan was to have a tour to get our bearings, and I knew that the boys would think of it as an adventure. We jumped into our carriages, Ben and me in one, Neil and Josh in the other, and for the first time it struck me quite how dangerous this mode of transport was. Here we were sitting in front of our driver, facing oncoming traffic, with no seat belts to secure us. I formed a human strap by putting my arm across Ben's lap and clutching the side of the rickshaw in a vice-like grip, while Neil did the same for Josh. Luckily, in 2005, Phnom Penh, had more rickshaws, bikes and motorbikes than cars or trucks on its roads, so although the traffic was dense it was also slow.

We passed run-down art deco mansions, originally painted in vibrant pinks, yellows and greens, which had long since faded and were in dire need of restoration. The city, though wonderfully atmospheric, was clinging by its fingertips to a more prosperous past, and it was the gleaming golden domes of the wats (Buddhist temple monasteries) that shone the brightest. The river front, where three rivers – the Mekong, Tonlé Bassac and Tonlé Sap meet – remained impressive, and we zipped along a motor-traffic-free esplanade lined with palms and poles with billowing flags. Our rickshaw drivers pretended to race each other to make the boys cheer, and when an elephant came swaggering into view on the road ahead Ben yelped in disbelief.

Our tour finished at Psar Thmei, the mustard-coloured art deco central market, where we were meeting a local guide named Munny, who would give us a food tour of the market and guide us through a local lunch. This was a cunning plan of mine to get the boys to be more adventurous food-wise. With a stranger in the picture it wouldn't be as easy to cry, 'I want pizza', and turn their noses up at trying some local nosh. Ben in particular had an aversion to anything remotely fishy, pegging his nose in an embarrassingly obvious way whenever we visited fish markets (although luckily this never failed to amuse the market traders). On a more serious note, as Josh had a nut allergy, I felt a little safer in having a translator with me, even with the trusted translation into the local language that I carried everywhere with me stating the seriousness of his condition.

Munny (a smiley chap in his early thirties, and quite formerly attired in a suit) was standing where we'd arranged to meet, at the entrance on the market's south side, and as we jumped off our rickshaws he lifted his left hand to his chest in the prayer position and bowed towards us. I noticed that his right jacket sleeve was tucked into the pocket, and that the bottom half of this arm was missing. I felt a slight panic as I wondered how long it would take Josh to cotton on. This was the moment I'd been dreading.

'I heard you English boys like eating spiders, right?' he asked the boys.

I'd filled the boys in on this particular Cambodian delicacy, reading aloud from the guidebook at bedtime the previous night to much pretend vomiting.

'No way,' said Ben, warming to the game quickly.

'How about grasshoppers?' Munny asked, sticking out his tongue.

'Yukky,' shouted Ben.

'Snake?'

'Ha! Ha!' chortled Ben.

'Okay, so you like chicken with lemongrass, right? Pork with noodles, okay?'

Ben looked at me for guidance.

'Sounds yummy,' I said. 'What do you think, Josh?'

Josh was unusually quiet.

Cambodians have a saying, 'For news of your heart, ask your face', and Josh's face at that moment was saying 'Help! I don't know where to look without getting into trouble.' I felt instinctively that Munny, with his jokey, open disposition, wouldn't be offended if I addressed right away what was so obviously on Josh's mind.

'Do you mind if I ask about your arm,' I ventured.

As I expected, Munny didn't bat an eyelid. 'No problem,' he said, patting the empty sleeve. 'I lost it when I was seven.'

'In a landmine accident?'

'Yes. I was in the fields near my village. I was playing a dice game called klah klok and my friend threw my dice away when I beat him. I was searching for them in a bush when the mine exploded.'

'I'm so sorry. That must have been awful,' I said.

'I am still here, and I am grateful for that; Munny said.

'Thanks very much for sharing that with us; I said, glancing at Josh, who was still looking everywhere other than at Munny.

Squatting down next to Josh, Munny asked if he was hungry. Josh shrugged but did, at least, make eye contact.

'I know best place for noodles; Munny told him. 'Then we eat mango ice cream, okay? Follow me.'

We snaked through the market, pausing often so that Munny could point out something unusual like a huge jackfruit that turned out to weigh around the same as Ben; bunches of hairy crimson rambutan (a cousin of the lychee); ansom chek (sticky rice balls filled with banana) and bowls of hopping frogs.

'Mummy, can I buy a frog?' Ben asked, peering into a white plastic bowl full of amphibians.

'Not today; I said, letting him, as usual, keep hold of a faint hope that on another day he may be more successful.

'Nicest spiced and fried; Munny whispered to me once Ben was out of ear shot.

In the centre of the market we reached a food stall circled with pink plastic chairs.

'Lunch stop; announced Munny, taking off his jacket. Now in short shirt sleeves, the stump of his right arm was visible, and I saw Josh's eyes widen. His chance to behave appropriately had arrived.

'Okay, boys,' Munny said. 'I'm going to order some food. If you don't like it no problem, but just give it a try.'

'Not spiders,' said Ben, looking suddenly worried.

'Haha! Spiders only for birthdays. Is it your birthday?'

Ben shook his head. 'Then, no spiders today, sorry,' Munny told him.

Meeting Munny was a gift – astute, kind and great with kids – by the end of our lunch Josh had lost all of his awkwardness and was happily listening to a story about a football team Munny had played for when he was at school.

'Can you really play football?' asked Josh.

'Yeah, sure. You don't kick with your arms, do you?'

Ben laughed at the thought of this, but Josh still wasn't sure.

'Can you ride a bike,' he wanted to know.

'Sure.'

'And roller skate?'

'Never tried.'

'Swim?'

'Not so good.'

Munny had earned the boys trust and they gobbled up all the dishes that the ladies on the food stall cooked up for us to sample, with Ben even taking a bite of fish amok (steamed fish with coconut wrapped in banana leaves), which was nothing short of a miracle.

I could have hugged Munny when we left him two hours later. 'Thank you. That was great,' I told him, nodding towards Josh, and I felt sure he understood that I wasn't just talking about the food tour he'd given us.

As we travelled around Cambodia, we couldn't have had a friendlier welcome from restaurant owners to tuk-tuk drivers to local guides, and all were charmed by our two young sons. The boys were at an age now where they enjoyed a little chat and loved to share a joke, and I relished the increased interaction we had with local people because of it. As a mother of boys, it was refreshing to hear a call of 'Lucky lady', rather than the usual 'Oh, two boys! Poor you!' that I got at home. Pregnant ladies furtively stroked me in crowds in the hope that some of my boy-producing hormones would rub off, but I never once felt that these caresses were inappropriate. By the time we reached Siem Reap, gateway to Angkor Wat, we were in love with the country and its warm-hearted people.

Hard to believe but Siem Reap was still quite sleepy back then. The building work that would turn the town into a tourist Mecca was in progress, but in early 2005 my first impressions were of a backwater. You didn't have to stray far out of the city to see

Cambodia as it had been for centuries where water buffaloes grazed at the edges of luminous green rice paddies and farmer's houses were built on stilts.

The hotel we were staying in had a brand-new kids' club, a rather surprising facility and no doubt there to cater for future Western guests. Spying all of the shiny new toys, the boys were keen to give it a go. Neil and I settled at the pool, excited at the prospect of this unexpected child-free hour. As is usual with stolen moments, the hour flew and I was soon back to collect the boys, unprepared for the scene of total carnage that would greet me. In the corner of the room, the young, inexperienced babysitter sat, looking like she might cry. Around her, the Lego box had been upturned, the home corner trashed, every jigsaw puzzle in the place scattered, every board game opened. The boys, on seeing me, looked like rabbits not just caught in headlights, but in megawatt floodlights. They both froze, halfway through a game of lob the dolly. I might have laughed, if I hadn't been so cross.

'What on earth?' I bellowed. 'Right, you're to pick up every toy. I want all the jigsaws back in their boxes, every bit of Lego off the floor, the books back on the shelves. Get to it, you naughty boys!'

The boys scrambled around.

'I'm so sorry,' I told the girl. 'They never normally behave like this. Why didn't you stop them?'

'I'm not allowed to,' she told me. I guessed that number one on her job description was that she should indulge her Western charges. If they wanted to run riot (and clearly they had) it was her job to let them.

'I'm very, very cross with you,' I told them again. Ben began to cry.

'And you can stop that,' I snapped. 'You haven't got time to cry.'

When the room was tidy, I stood the boys in front of the babysitter. 'Say sorry to the kind lady!'

'Sorry,' they both chimed, looking sheepish.

'Right, let's find Daddy and tell him what's been going on.'

Annoyingly, I could see it was an effort for Neil to look angry. He was in that happy late-afternoon holiday state, post three beers, pre-doze, as I filled him in on events.

'Very naughty but what's the saying, give an inch... Anyway, you're never ever to behave like that again, do you understand?' Neil asked the boys, managing to conjure up a frown.

The boys stared at their feet and nodded.

'Just out of interest though, what was it like being allowed to be that naughty?' he asked.

Josh looked at Ben and giggled. 'Really fun,' he said. Josh was nothing if not honest.

I wondered if the breathtaking splendour of Angkor Wat, originally constructed as a Hindu temple complex in the beginning of twelfth century, and converted to Buddhist at the end, might be lost on the boys, but for the three- and four-year-old who had recently watched carefully selected scenes from *Lara Croft: Tomb Raider*, which was filmed on location there, it was like visiting a movie set.[7] 'Is it real?' Josh kept asking, in obvious awe of the immense stone structures, many still yet to be stripped of the vines and jungle foliage that had camouflaged them for centuries. We walked along a terrace of stone elephants, but quickly detoured off in search of shade as it was stiflingly hot. Glimpses of the temples' former glory were found in beautiful relief sculptures carved into the stone, many of which you could only view once you'd pulled away the ferns and creepers that covered the walls. It was like walking through an enchanted forest; our steps were cushioned by soft lichen and moss, and soaring banyan trees with twisted buttresses had grown feral, bursting through temple roofs. This was a place to let your imagination run free. The boys were explorers like Indiana Jones one minute, soldiers on dangerous manoeuvres the next. Dinosaurs lurked around every corner. The strangeness and scale of the place inspired a succession of games, until they were exhausted.

---

7 *Lara Croft: Tomb Raider* is a 2001 action-adventure film based on the Tomb raider video game series.

'Snack time?' I suggested, just as the heavens opened and the rain came down. We ran around like loons, jumping in and out of the quickly forming puddles, with the boys shrieking at every crash of thunder.

When the storm passed, we took off our shoes and waded back along the path that was now running like a stream, to the steps of a minor temple, where we sat to dry in the sun and refuel on wafer biscuits and water. In the rising heat, the smell of earthy vegetation engulfed us, and steam rose off the ground like mist. At the centre of the temple's inner courtyard, a skinny shaven-headed boy stood in the mud. Perhaps only a few years older than Josh, his job was to sweep the floor with a broom that was the same size as himself, but the sudden rainfall had made it impossible. He paused in his drudgery to check out Josh and Ben shyly, then slowly wandered towards us until he was around 6 metres away. Only then did I see that his left arm ended in a stump. My heart ached imagining how hard his life must be.

'I think he's looking at our snack,' said Josh.

'Do you think he'd like one?' I asked.

Josh nodded.

'I come too?' asked Ben.

'Okay, but let Josh give the boy the biscuit,' I said, handing the wrapped wafer to Josh.

I watched my four-year-old son walk towards the boy trailed by his little brother, and knew he'd learned his lesson. There'd been no pointing, no staring, no disrespect for this boy's feelings. Josh handed the snack to the boy, who tucked it into the top pocket of his shirt, two sizes too big. It was plain that he was saving it as a treat. He nodded and smiled at Josh, as if to say well done.

5

Borneo

A Lesson in Perseverance

Josh age seven, Ben five and Freddie two

28 December 2007,
Sandakan, Sabah

For dinner with a difference tonight we headed to The Ocean King Seafood Restaurant, a vast place built on stilts above the water in Sandakan.

'You're joking!' exclaimed Josh, when I broke it to him that the pools teeming with fish and crustaceans that we were staring into were in fact the menu.

'I thought we were in an aquarium,' he said.

Ben peered into the tanks where strange prehistoric creatures such as horseshoe crabs, sea cucumbers and spiky sea urchins bobbed. 'Nah,' he said. 'Silly, Mummy. We're not going to eat those.'

With no menus, it was a simple point and nod that got you what you wanted, which in Ben's case was nothing what-so-ever.

Once Josh had recovered from the shock, he chose the least exotic-looking fish he could find, which I guessed to be some sort of snapper. Freddie, still thinking it was a game, chose an iridescent blue-striped fish, far too pretty to eat.

'You can't eat that. That's like eating Dory,' cried Ben.[8]

'You know, I've heard that the prettiest fish don't taste the best,' I told Freddie. 'Let's save those to snorkel with, shall we?'

The snapper was delicious, steamed with ginger and garlic. Josh and Freddie gobbled it up, while Ben ate only fried rice.

'You want dessert?' asked the waitress.

I resisted the urge to suggest jelly . . . fish, as one look at Ben's grumpy face told me that this was no time for making 'mum jokes'.

~~~

'You're going where?' asked my Dad.

'Borneo,' I repeated, guessing that he was hoping that he'd heard incorrectly, and I'd said Bournemouth, Bognor or Barry.

He glanced at his precious grandsons. Josh aged seven, Ben five and Freddie a whisper off two.

'Is that a good idea?' he wanted to know, images of head-hunting tribes no doubt marauding through his imagination.

---

[8] Dory is a blue tang, one of the main characters in the 2003 computer-animated film, *Finding Nemo*, created by Pixar Animation Studios and released by Walt Disney Pictures.

'Can't think of a better one,' I told him.

'But is it safe there?' he continued.

'We'll be fine. And by the way I've got three words to say to you on the subject of safe holidays: shotgun, farmer, France. Never mind riding pillion on a motorbike without a helmet.'

'Those were different times,' he said.

'If you say so. And anyway, I'm writing about it, so it's partly for work.' As always, this was the trump card I could always rely on.

'Just promise me you'll be careful,' he said, accepting that it was useless to argue further.

I didn't admit it, but I thought that my dad might have a point. Was this such a good idea? This was the first long-haul adventure trip we'd taken with three children, and in choosing Borneo we were certainly throwing ourselves in at the deep end. In Thailand and Cambodia we'd each had one child to keep an eye on, to pick up and carry, to sit with on planes. Five was such an awkward number in so many ways. Now there'd be three curious, and, at times, overly confident boys to keep track of in a destination that was unfamiliar and full of potential hazards. The run-up to this trip had been particularly busy and beyond reading *Where The Wild Things Are* (the closest thing we had at home to a children's book on Borneo) and making sure our immunizations were up to date, I'd done little to prepare the boys.

For the first time in my life I didn't sleep the night before our departure, sick with sudden worry.

The day after Freddie turned two, we flew to Kuala Lumpur, the gateway to our adventures in Sabah, Borneo. Our plan was to let the kids acclimatize and shake off their jet lag while staying in the Traders Hotel with its views to Petronas Towers. We'd been there just two hours when, while paddling in the shallow end of the rooftop swimming pool, Ben suddenly screamed and clutched at his foot. He'd stepped on a diamond stud earring and it had pierced deep into the sole of his soft skin. Of all the snake-biting, kid-stealing, boat-sinking disasters I'd envisioned over the last few days, this wasn't one of them and strangely, after pulling out the earring (a trophy, which Ben was reluctant to give up), I relaxed. It was impossible to predict what lay before us but one thing was for sure, anticipating danger at every turn would only ruin the trip.

Kota Kinabalu is the capital of Sabah, Borneo's most northerly state, and it was shrouded in low brooding cloud the colour of granite when we arrived after a two-and-a-half-hour flight from Kuala Lumpur. Humidity was high and our driver warned that a storm was brewing, and that we should hurry to our transport. Ben, still milking his recent impalement, limped (on the wrong foot) from the surprisingly modern terminal to the old banger of a minibus I'd pre-booked to pick us up. Our destination, Shangri-La's Rasa Ria Resort, lay between rainforest and sea, an hour's drive north and I watched as the wooden crucifix, that hung from the rear-view mirror, swung at every bump and twist in the pot-holed roads. As our driver had predicted, the heavens opened just ten minutes into the journey.

'I can't shut the window,' I shouted back to Neil who was at the back of the bus. In fact, none of the windows on the right side of the bus would budge so we all scooted to the left and watched as the raindrops bounced on the seats next to us.

'Nice! Refreshing!' our driver bellowed above the clatter of rain on the roof. No doubt he was hoping the fact we were getting soaked to the skin wouldn't ruin his chance of a tip.

There were two reasons to bed down at the Rasa Ria for Christmas – its stunning location and its private nature reserve, home to eight young Bornean orangutan orphans. This was to be our first orangutan encounter and the boys were both excited and a little nervous.

'Will the monkey cuddle me?' Josh wanted to know.

'Orangutan,' I corrected him.

'Yes, but will the monkey come into our bedroom?'

'Orangutan. And no, they won't cuddle you or come into your bedroom.'

Josh looked relieved. By now the sun was out and we headed for the reserve – actually more a halfway house for waifs and strays, at liberty to roam during the day and play among the vines and vegetation, but by night housed and protected by rangers. Their next residence would be the famous Sepilok Rehabilitation Centre, before eventually being reintroduced to the wild. I held Freddie monkey-style on my hip to give him a better view, and soon the young orangutans spotted him, and hurried over to check out his mop of scruffy fair hair and cheeky grin. He entertained them with a series of impressive and well-rehearsed 'oo-ooos'.

'Do it again, Freddie,' instructed Josh, impressed by his baby brother's linguistic skills. Ever the performer, Freddie obliged and then stuck out his tongue at a five-year old orangutan, named Cute, who stuck his right back out at Freddie. We all squealed with delight.

'Again, Freddie,' begged Ben, jumping up and down. If you were ever looking for a wonderful example of the gift children have in simple communication, then this was it. What fun these two were having without the need of a language, and what a pity that we seem to lose this ability as adults. Freddie and Cute had bonded that was clear, and sharing ninety-seven per cent of the same DNA why wouldn't they? They stuck their tongues out at each other a few more times before a piece of sugarcane caught Cute's eye and he ambled off.

On our way back to the hotel, via the beach, Ben spotted a four-foot-long monitor lizard sauntering along the sand, which prompted a game of Jurassic Park.[9] There'd been a breakdown in the park's security system, and they had to fix it before the monitor lizard ate them alive; the kind of game that involved a lot of running around in circles and screaming. The monitor lizard, meanwhile, paid them no attention at all, waddling off into the bushes.

Lovely as the resort was, I insisted that each day we take an excursion. At the biggest *tamu* (market) in Sabah, we wandered past mangoes the size of melons, spiky foul-smelling durian and dried fish that were piled into mounds, but it was the stalls selling musical instruments that caught the boys' attentions. While Josh and Freddie had a go at banging on a type of wooden glockenspiel, I noticed Ben whispering to Neil, and I was told to look away while he made a purchase.

'You can look now,' Ben said, handing me a small metal gong.

'It's for bashing at teatime,' he told me, looking pleased with himself. Ah, the daily holler to get the

---

[9] *Jurassic Park* is the 1993 Sci-fi/Action film in which cloned dinosaurs, held in a wildlife park, escape and go on the rampage.

boys to the kitchen to eat their tea. It drove me insane, as my sons well knew.

'Oh, I love it,' I told him. As gifts go, it remains one of the best I've ever received.

On Christmas Eve I arranged for a taxi to take us to a nearby jetty, where, according to the guidebook, we could catch a local boat upriver to visit a sea gypsy village built on stilts above the water. The place ticked lots of boxes for me – educational, unspoiled and a glimpse into local life with no other tourists around.

'Why you want to go? No one goes there,' asked the bemused concierge.

'Exactly,' I replied cheerfully, relishing the idea of being off grid for a few hours, far away from the piped Christmas tunes such as 'Santa Claus is Coming to Town' that were being played on a loop around the resort. However, the boys, who were happily playing in the swimming pool, thought it sounded like a bad idea. Even Neil, who was usually game for any adventure, was reluctant to peel himself away from his sun-lounger.

'But it's Christmas Eve,' Ben pointed out. He was starting to worry that Father Christmas wouldn't find us (even though, with my help, he'd written to the big man with clear directions, including a map); and all Josh could talk about was how Father Christmas was bringing him a Nintendo Wii. He wasn't. Oh, dear.

'We'll only be out for a few hours,' I coaxed. 'And remember, we didn't come all this way to lie by a pool.'

After promising that we would be back in time for the Christmas dance show, we set off.

The taxi driver dropped us at the jetty as I'd instructed, where a large dug-out canoe was just about to depart, already full to capacity. After much shuffling space was made and we boarded, handing over the equivalent of one pound fifty for the five of us.

'See, this is fun,' I said, just as the first spot of rain landed squarely on my forehead.

The local people quickly unfurled scraps of plastic sheeting to shield from what was now turning into a full-on rainstorm. My neighbour, a lady with an ample bosom, reached over and scooped Freddie up, plonking him on her lap out of the rain. It was obvious that she considered Freddie to be in a safer pair of hands with her. Likewise, room was made for Josh and Ben under cover, while Neil and I were left to get wet.

Once at the village, Freddie's self-appointed guardian gestured for us to follow her, and like a straggle of wet ducklings we did as we were told, over slippery wooden walkways to her house on stilts. We watched as she put a pot on a small wood-burning stove to boil water to make a weak tea that she then served to us in tin cups. Then she gestured that we should sit on the floor.

'What now?' whispered Neil.

'Drink your tea,' I hissed.

'You've got to be bloody joking. I don't want dysentery for Christmas.'

In my cup a few indistinct leaves floated, and I pretended to sip the tea to be polite. The boys were unusually quiet, all with eyes on me. 'Well, isn't this nice?' I ventured.

'People are staring at us through the window,' said Ben, which was true. In fact, over the course of the next hour, the whole village trooped by to take a look at us through our host's window, which wasn't difficult as there was not one pain of glass. Word had obviously gone around about the idiot foreigners who had taken their children on a crazy excursion in a storm, and while we waved and smiled at everyone, our benefactor looked proprietarily on, like the cat who'd caught the canaries. 'We'll make it back for Christmas, won't we?' sniffed Ben, staring out to where kids his age were beetling around in small canoes in the rain.

'Of course, we will,' I told him, sounding surer than I felt.

'Because there's no way Father Christmas will find us here,' he added.

'And if he did, I couldn't play with my new Wii,' added Josh.

It was true (and not just because he wasn't getting one). As far as I could see there was no electricity. On my travels, I'd been to other remote places, where the indigenous people had half a Currys warehouse packed into their tents/huts/shacks.[10] Here, there wasn't a TV or satellite dish in sight.

'Don't worry as soon as the rain stops, we'll go. Won't be long now,' I told them.

'You promised we'd be back in time for the show,' said Josh.

'Hmmmmm,' I replied. 'Well, fingers crossed.' Missing, what was sure to be some God-awful show, was at least the silver lining to this particular black cloud.

After an hour, the storm began to ease, and the sun broke through. I pointed back to the jetty and our host got the message, leading us back to the pontoon where a return journey was secured at three times the price of our passage up stream (fair dos). I tried to give her a tip for her kindness, which she refused, choosing instead to have one last cuddle with Freddie. When I look back on that day, it's not really the rain, which soaked us to the bones, that I remember, but rather the kindness of the stranger who took us in. Josh needn't have worried, we were

---

[10] Currys PC World is a British electrical retailer, founded as 'Currys' in 1884.

back in plenty of time for the Christmas Eve buffet and show (damn it), where we struck it lucky with a table next to the chocolate fountain.

'Is that what I think it is?' asked Neil, pointing to where a porcupine was casually wandering on to the terrace.

The waiter flipped a napkin towards to creature in an attempt to shoo it away. 'It's looking for salt,' he told us.

'Wouldn't it prefer chocolate?' asked Ben, twirling his recently dipped marshmallow.

'Oh no,' said the waiter. 'Porcupines prefer salty things; leftover dried fish are their favourite.'

'Yuk,' said my fish-hating son.

As if to prove the waiter wrong, the creature flicked the floor with its long tongue, licking up the trail of chocolate left by the boys as they had ferried their marshmallows on sticks back and forth. 'Ha!' said Ben smugly.

When Christmas was over (Father Christmas did find us and arrived in a helicopter – an event so mind-blowing that Josh forgot all about the Nintendo Wii), we travelled on to Sandakan on the East coast. The attraction here was the aforementioned Sepilok Orangutan Rehabilitation Centre, founded in 1964 and home to around eighty orangutans, which sprawls over 43 square kilometres of protected land in the Kabili-Sepilok Forest Reserve.

Our home for the next few days was a basic wooden chalet at the Sepilok Nature Resort, built among a rainforest foliage of palms, pitcher plants and giant ferns, and near enough to the rehabilitation centre that we could hear the orangutan's soft hooting night-time calls. With their new Christmas binoculars glued to their eyes, Josh and Ben went off to explore gardens alive with croaking toads, over-excited grasshoppers and rowdy indigo flycatchers, however soon came running back when a frog leapt on to Josh's shoulder. We were excited to find out if Freddie, the orangutan whisperer, still had the gift and spent the next two days visiting the rehabilitation centre to watch him at work. He chatted away enthusiastically to any orangutan that would listen and of the five of us, it was always Freddie who caught their eye, but he never did find another quite as obligingly responsive as Cute. These were happy hours spent observing the energetic apes as they somersaulted and swung along the jungle vines, and we agonized over which orangutan to adopt, finally agreeing on a young adult male called Sen. Twenty-five pounds would keep him in food and shelter for another year and we were promised a letter every three months from our surrogate primate. 'He's cleverer than you, Freddie,' I heard Ben whisper. 'You can't even write.'

In the evening we left the boys squashed into one bed after they'd clapped eyes on a spider with a body the size of a jammy dodger biscuit (Ben's rather unique analogy, not mine), while Neil and I sat on the veranda to make plans for the remainder of our stay. To have come all this way and not to have seen an orangutan in the wild seemed ridiculous. And yet, we knew the chances were heartbreakingly slim. Illegal logging, mothers with babies caught for the meat and pet trade, and the slash and burn technique often used to make way for palm oil plantations had rendered the Bornean orangutans endangered. Perhaps it was already too late. Plus, we'd been told that travelling on from here with young children wasn't advisable. I'd heard that orangutans were sometimes spotted near to the Kinabatangan River, Malaysia's longest and rich with wildlife, which was a mere 25 kilometres away, but the road to get there would be rough. Fleetingly, I thought of my dad's concern over this trip, and then voiced what I knew Neil was already thinking. 'Let's give it a go,' I said. We arranged a car and driver to pick us up a day later.

Our Malay driver and guide, Adam, raised his eyes at the sight of our young sons, then quickly rearranged his face to a broad smile, high fiving each of them as they climbed into his 4WD truck.

'Woo hoo!' whooped the boys. I watched them happily bouncing about, their bums rarely making contact with the seats as we bumped over potholes and through ditches along the dirt-track roads. Oh, the joy of being able to stay that relaxed while being flung about like rag dolls. Unlike me. I was gritting my teeth with every bone-crunching jolt to my spine. When we reached the jetty, we transferred to a small motorboat and Adam told us to keep a keen eye out for shy pygmy elephants and also for proboscis monkeys, which with their enormous great hooters, protruding potbellies and permanently erect penises, were soon providing fodder for my three's school-boy humour.

'And orangutans?' I asked.

'Unlikely,' came his reply.

'But perhaps, if we're lucky?'

Adam shrugged, reluctant to make promises, pointing to a brilliant blue stork-billed kingfisher that was flitting along on the bank. Pig-tailed macaques gave aerial displays, leaping from tree to tree and pied hornbills bellowed for attention above the constant shouts of, 'Look at his willy!' from the boys each time a proboscis monkey came into view.

At every rustle in the trees my heart missed a beat as I scoured the treetops for a flash of that distinctive ginger coat, but the orangutan was either elusive or never actually there. Either way, I resigned myself to the probability that we'd never see one in the wild.

'It's been a great day,' I told Adam, when we were back at the jetty. It was true. It had been. However,

I couldn't help but feel a little disappointed. Freddie had fallen asleep, and the other two were flagging and making requests for treats we had no hope of producing like ice creams and sweets.

'Do you want to go here on the way back?' Adam asked us, pointing to a crease in a well-worn map.

I gestured at the sleepy boys. 'Perhaps it's too much for them.'

He was suggesting a stop at the Gomantong Forest Reserve, famous for its caves where swift's nests are harvested for the Chinese delicacy of bird's nest soup.

'Amazing at sunset when the swifts return,' he promised me.

Again, Neil and I reminded each other that we wouldn't be passing by here any time soon and so we agreed. The caves weren't much to look at from the outside, and even before we'd set foot in them, we could smell the toxic stench from the accumulated excrement of swifts and bats, known as guano. We ventured to the mouth of the cavern and I looked at Neil. 'What do you think?' I asked.

'I don't like it,' Ben said. 'Please can we go back?'

'It stinks,' sated Josh, who wasn't known for mincing words.

Neil shrugged and picked up Freddie to carry. 'We're here now,' he said, taking his first step in on to the slippery guano splattered floor. 'Come on, boys! We can do this!'

Adam assured us that we would be through the cave and out the other side in just ten minutes. It was fetid and gloomy, but unfortunately not quite dark enough that we couldn't see cockroaches and giant millipedes scurrying around; I had to stop myself shrieking.

'You're doing great,' I told them, just as Ben slipped and put his hand down to save himself.

'My hand's got poo on it,' he wailed. 'I want to go back to the car.'

'Oh God, this was a terrible idea,' I conceded, trying to wipe the odious substance off Ben's hand with a tissue. Neil handed Freddie to me and scooped Ben up. Josh, as the oldest, had to fend for himself and we soldiered on.

'Peg your noses,' I barked, breathing through my mouth.

After ten minutes of picking our way through the cave, a large shaft of light came into view and we hurried towards it. 'Sorry, boys. Never again. We'll take you to the Isle of Wight next year,' I promised.

Adam had gone ahead to clear the path by kicking as much debris out of our way as he could. 'Nearly there,' he called back to us. Up ahead I could see daylight (our exit from, what I was secretly calling in my head, 'this God forsaken shithole'). As we stumbled out of the cave, into a clearing in the forest, I couldn't have felt guiltier at having put my young sons through such an unpleasant ordeal.

'Oh, boys. I'm so sorry. That was awful,' I said. 'We made a bad decision, but I'll make it up to you . . .'

Adam gripped my arm. 'There, look there,' he whispered.

I gazed up to where he pointed, a sudden movement catching my eye. My breath caught and my heart seemed to thud to a stop as I realized what was before us. 'Look boys. Oh, my goodness, just look.'

Up in the bamboo, right before us, was the world's largest tree climbing mammal – a female orangutan reaching up for the newest, sweetest bamboo leaves with her long grasping fingers. No one said a word. Even Freddie was silent, lost for his usual 'oo-ooos'. Her shaggy reddish fur was aflame in the setting sun. So intent on feeding, she wasn't disturbed by our arrival and so we crept a little closer. Close enough to smell her musky odour, and to see how her feet gripped the bamboo stem, while she stretched her seemingly elastic arms up through the canopy. Then with a sudden movement she disappeared into the forest, leaving the trees quaking in her wake.

'Did you know she would be here?' Josh asked me.

I shook my head. 'No idea. Aren't we lucky?'

Adam agreed. 'I've only ever seen one other so near to the cave.'

I was close to tears as what we'd witnessed sank in. 'We almost gave up,' I reminded the boys. 'See how it was worth it? We would have missed seeing her if we'd turned back.'

Ben sniffed suspiciously at his hands, and yet even he nodded.

To have seen the original wild man of Borneo not on a feeding platform but in a tree, feral and free, remains one of the rarest privileges I've had while travelling, made all the more special by sharing it with my family.

6

Egypt

A Lesson in Moral High Ground

Josh age nine, Ben seven and Freddie four

27 December 2009,
Valley of the Kings, Luxor, Egypt

We were up with the sunrise today, just as the black ibis were flying in a perfect 'V' shape down the centre of the Nile on their way to breakfast on potatoes in nearby fields. My plan was to get to The Valley of the Kings early, and to pay our respects at Tutankhamun's tomb before the dreaded tour buses arrived. The fact that Tutankhamun was a king at nine years old (the same age as Josh) has really captured the boys' imaginations, and they were unusually quiet as we made our way into the dimly lit chamber where King Tut has rested for three thousand three hundred years. Freddie held on tight to my hand as we filed past the quartzite sarcophagus to peer in at the ornate wooden coffin that holds his mummified body. The room was dank, the air stale, and the wall paintings that depict his life and death spotted with marks left by fungus and mould – a result of the dust and humidity that comes with allowing up to one thousand tourists to troop through daily. It was shocking to see how a place of such historical importance had been left to rot and the boys looked under-whelmed.[11] This was a far cry from the splendours of the Tutankhamun exhibition we'd seen in London, and we emerged back into daylight feeling gloomy. I was scouring the map of the tombs and deciding which king or queen to call in on next, when a shifty looking guy carrying a bucket (most likely some kind of caretaker) sidled up to us.

'Want to see the shop of the hairdresser?' he asked.

Neil and the boys looked to me. Normally in a situation such as this, I'd suggest politely that he should be on his way, but the idea amused me. Could there really be such a thing? It wasn't something that was marked on our map, and it was such a ridiculously ordinary thing to be offered the chance to view. What we needed right now was a little light entertainment, so I agreed to follow him.

'Did he cut Tutankhamun's hair?' Ben wanted to know, already creating a back-story.

The man smiled, while he weighed up just how gullible we may be.

'Of course!' he said, glancing from side to side in a satisfyingly villainous way. 'Follow me.'

---

11 Since 2009, things have improved. Just after our visit, a ten-year restoration project began, to repair the ravages of mass tourism and install a new air filtration and ventilation system to protect the tomb.

So, we did. Off the main tourist trail we went, to the very back lot of the excavations, where work had either been abandoned or was only just beginning, to an immense oblong pit, with rough-hewn steps to its bottom. I could make out a few doorways, a doorstep or two, of what might have once been a row of shops. Fully immersed in the adventure, the boys were keen to scramble down, but I stopped them. We all peered into the hollow.

'Look there!,' instructed our guide, pointing to a plinth above one of the doorways. Carved into the stone, was what looked like a pair of scissors.

'He's done that himself,' whispered Neil. 'He must think we were born yesterday. What's he going to show us next, the tanning salon next door?'

On a whim, I've just typed 'did the Egyptians use scissors' into Google, and this is what came up: 'Egyptians created a scissor-like device around 1500 BC, consisting of two blades connected by a spring-like mechanism'. Well, who would've thought it? Egyptian wonders really will never cease.

～

No need to rally the boys' enthusiasm for a trip to Egypt. Ever since we'd taken them to see the Tutankhamun exhibition at the 02 Arena in London, in all of its shiny golden glory, our sons had been dazzled by the Egyptians. Subsequent trips to The British Museum, plus DreamWorks' animation

The Prince of Egypt, combined with Scooby Doo's *Where's My Mummy?* had further fuelled the obsession. Of the three, Ben was the biggest wannabee Egyptologist and I'd managed to find him a kid's guide to hieroglyphics, which he packed in a self-styled archaeologist's knapsack along with a torch (for seeing in tombs), a pencil and notebook (for recording any new discoveries), and his teddy, Spencer, (after all, every great archaeologist needs an assistant).

As our flight was at midday on Boxing Day, I'd packed before Christmas. All that remained to do was to cram in our wash bags on the morning of our departure but that was proving difficult. Since I'd packed, several impractical Christmas presents had been smuggled into the bags, including a Bop It! (a really annoying audio game that issues a series of commands, like 'Bop it to start!', over and over again), a Star Wars Lego set, a basketball hoop and a barbecue play set, including five plastic sausages (one each, Freddie later explained). Of these, only one sausage made it to Egypt as a stowaway. Burned out after the usual mania of Christmas with three young kids, making Cairo our first stop suddenly felt like a bad decision. What we needed was some down time, and after discovering that two of our bags had gone missing and that the driver I'd booked to pick us up at the airport had made other plans,

I sat in a taxi and stared despondently at Cairo's unattractive ring road. All I could imagine was the beach at Sharm-el-Sheikh that Neil had suggested starting the holiday with, which I'd vetoed, insisting that we stick with our tried and tested formula for a successful family holiday, namely culture first and beach second. As a Christmas treat, we'd splashed out to stay at the Marriott Mena House hotel, originally built as a royal hunting lodge in 1869. It was the closest hotel to the Great Pyramid of Giza and the Sphinx, and I reminded myself that once we arrived, all would look better.

'Mum, where's the pyramid?' Ben asked. 'I can't see it.'

From the window of our pyramid-view suite, we all gazed out into the pitch dark. He was right. It wasn't there.

'I'll ring down to check if we're in the right room,' I said, irritated that something else was not going according to plan.

Like a scene from *Fawlty Towers*, I interrogated the front of house manager regarding the location of the pyramid.[12]

'I booked a pyramid view,' I told him.

---

[12] *Fawlty Towers* was a British sitcom that was broadcast in 1975 and 1979. The main character, Basil Fawlty (played by John Cleese), was the belligerent owner of a hotel in Torquay called Fawlty Towers.

'And you have one, Madam.'

'Where is it then?'

'Directly in front.'

'I can't see it.'

'But Madam, it's dark.'

'Yes, but surely we'd still see it. It's massive.' I didn't mention that I'd expected it to be illuminated by night, a la Blackpool Tower.

'Not by night, Madam.'

He was calling me 'madam' a little too often for my liking, but Neil later said that he had hit the nail right on the head with that particular form of address.

'It's a bit disappointing to be honest,' I said.

'I can only apologize.'

'Hmph, well okay then. I'll just have to take your word for it.'

The following morning, I felt such a fool. When I drew back the curtains my nose was practically pressed up against the Great Pyramid, and Neil enjoyed himself by doing a Basil Fawlty impression, shouting 'Madam, what did you expect to see out of Torquay bedroom window? The Great Pyramid of Giza?', while the boys laughed their heads off, without even understanding the joke. Serves me right.

Although the Great Pyramid was indeed on our doorstep, we decided to make the Egyptian Museum in central Cairo our first stop of the day. That way we could view what wonders the now empty pyramid once held before exploring it.

'How come the Egyptians built the pyramids so close to the shops?' Freddie wanted to know. To be honest, even I had imagined them a little more out in the desert.

'The shops weren't here when the pyramids were built four and half thousand years ago,' I explained, conscious that at Freddie's age last week must feel like a long time ago.

Over the last century Cairo had slid towards them like a dirty oil slick, making the effects of pollution on these triangular edifices the number one concern for archaeologists. This was all too visible on our journey by taxi from the suburb of Giza, over the Nile River, and into the traffic-clogged centre of the city and a grey haze clouded the air, visible even in the shimmering mirages that lay before us on the road. Grimy though it may be, Cairo is still an alluring city and the boys were wide-eyed with it all, from seeing a man pushing a cart of pomegranates up the middle of the Egyptian equivalent of the M25, to the camels lounging by the roadside, to the hustle and bustle of street markets from which we caught a waft of roasting nuts and spit-roasted lamb shawarma through the air-conditioning. In some areas of the city, Islamic architecture dominates in the shape of minarets and domes, in others there are warrens of centuries-old streets dating from Roman times, but in the city centre, it's nineteenth-century European architecture that's the backdrop to Middle Eastern street life. It felt exciting to be in the midst of it all.

Home to Egypt's Pharaonic antiquities since 1902, the Egyptian Museum is a dinosaur and, in 2009, on the edge of extinction.[13] On the one hand, I rather liked the shafts of dust-filled light and the scurrying cockroaches that darted quickly across the stone-slabbed floors. They reminded me of the documentaries we had watched about archaeologists working in Cairo in the 1920s, conjuring up romantic images of dapper men in pith helmets and plus fours, smoking pipes. On the other, I couldn't believe how carelessly the priceless treasures were displayed and cared for, and my blood soon began to boil.

At the entrance to the museum, three small children around Josh, Ben and Freddie's ages climbed onto the back of a sculpture of a sphinx, and understandably, because it looked fun, Freddie was all for joining in. 'This isn't a playground,' I told him. 'That monument is thousands of years old.'

---

[13] Three years after our visit, construction began for a new state-of-the-art archaeological museum – the National Museum of Egyptian Civilization – which will supersede the old Egyptian Museum.

'But their Daddy's letting them.'

'Yes, well their Daddy should know better,' I answered very loudly, glaring at the father who was busy taking photos of his kids riding the sphinx like a bucking bronco.

Inside it was no better. While bored museum attendants looked on, tourists laid their sticky fingers upon every ancient Egyptian wonder they could get their hands on. I observed in horror as a limestone statue of Queen Nofret had her breast stroked by a passer-by, and the head of Queen Nefertiti, dating from 1340 BC, was also casually pawed over.

'This place is unbelievable,' I muttered crossly, watching a tourist hoist himself upon the base of a monumental statue of King Djoser.

'Is that allowed?' I asked one attendant, as another visitor posed for a photo, hoisting up her skirts to straddle Horus, the falcon-headed god.

'No English,' he told me.

'Mum,' hissed Josh. 'Don't make a fuss.'

'Someone has to,' I told him.

I'd experienced this kind of disregard for cultural heritage in other places. At Carthage in Tunisia, I'd met a man who was collecting (stealing) pieces of masonry from the site, in full view of guards who I suspected he had paid to turn a blind eye. In Fatehpur Sikri in India, I'd got into an argument with a tourist whom I caught mid deal, buying antique coins from a cleaner. I'd heard, that in Cairo, 'Looking to buy something special? Genuine artefacts? was also a frequent offer from real-life tomb raiders. The sad reality was that where there were tourists with money to burn and questionable ethics, there would always be a black market in antiquities.

There were situations when Neil was prone to forgetting that we had three young children in tow and would revert to his habits of travelling pre-kids and visiting museums was one of them. With gritted teeth, I watched him read every label on every exhibit, often back tracking to reread a notice for the second or third time, oblivious to what was going on around him. Meanwhile, I fended off a steady stream of dodgy-looking geezers, who popped out from behind pillars and statues to offer their services. 'You want unofficial guide?' was the usual question, hissed out of the corner of their mouths.

I'd been warned to take the official tour or go it alone. Cairo was full of chancers with no knowledge of Egyptology. It was at this moment that Ben began to earn his keep.

'No thanks,' he informed them cheerily, holding up his guide to hieroglyphics and the torch, which moments before he'd been shining into a sarcophagus to illuminate the remains of a mummy.

'I bet he's saying, "Darn, that pesky kid!",' I joked as we watched each one skulk off. I was referencing what the baddies in Scooby Doo were prone to mutter once Scooby and 'the gang' had foiled their dastardly plans and it was a comparison that Ben liked. For the remainder of our trip (unless we took an official tour) he stepped swiftly in to save us from unwanted solicitations.

Meanwhile, Freddie was grappling with the concept that what he was seeing was 'real'.

'Is that a real mummy?'

'Yes.'

'And is the mummy a real person?'

'Yes.'

'And was that person a real Egyptian?'

'Yes.'

'And were the Egyptians really real?'

'Yes.'

By now I was beginning to realize that going down the Prince of Egypt/Scooby Doo route of introducing the Egyptians to Freddie probably wasn't my best idea. The reality of these often quite gruesome remains bore no resemblance to the comically bandaged mummies in cartoons. On peering into one coffin, we found a gnarly knuckled embalmed hand protruding from the cloth, in another a shrunken skull. Suddenly the Egyptians didn't look quite so glamorous and the suitability of bringing a four-year-old to see such horrors was a questionable decision on our part.

'Bit late now,' Neil said, when I voiced my concerns, which was what he always said when I was futilely worrying over things that it was too late to do anything about. I watched him wander back to reread the caption on a sarcophagus that he'd already spent ten minutes staring at.

'Enjoying yourself?' I asked.

'Lovely time. Thanks,' he replied.

'And is that real?' asked Freddie, pointing to a wizened body with skin like crispy bacon that looked more like an extra-terrestrial than a human being.

'No. Not that one,' I lied. 'Come on, let's go and look for Tutankhamun's golden burial mask.'

Back on the Giza Plateau we fended off proposals for camel and horse rides; ignored the boys' requests to stop at souvenir stalls selling tacky gold pyramids, sphinx keychains, papyrus scrolls and stuffed camels; shook off yet more unofficial guides; gave a snake-charmer a wide berth; slipped by the touts selling knock-off sound-and-light show tickets; and finally arrived at the entrance to The Great Pyramid. Built around 2550 BC by Pharaoh Khufu, it's a spectacular edifice, made with over two million blocks of stone, each weighing between 2.5 and 15 tons, and we were going inside. As I watched Neil buy the tickets, my mouth ran dry. I don't like small spaces. I try not to go in lifts. In a theatre I always book seats by the aisle. In hotels I check where the emergency exits are. I don't often admit to suffering from claustrophobia, but it's there, in my life – the suffocating, breathless horror of being trapped. So, the idea of entering this massive block of stone and walking down narrow airless dimly lit passages to the chamber at its core was quite simply a nightmare. However, my desire to share the experience as a family, combined with a dogged resolution that I didn't want the boys to grow up witnessing an irrational fear stopping me going about my life, spurred me on. With Freddie's sweaty little hand clasped in mine (for my comfort not his), I stepped into the dingy tunnel.

'You okay?' asked Neil, no doubt braced for a potential pyramid meltdown. Inside, it was worse than I'd imagined with the air hot, thick and still.

'Not sure,' I admitted. 'Let's just get it over with.'

Neil and I were forced to make the ascent hunched over as the tunnel wasn't large enough for us to walk upright in.

'Oh, God. This is awful,' I muttered, feeling the sweat trickle down my back.

'Mummy, you're hurting my hand,' Freddie said. I rather reluctantly loosened my grip and he pulled me up the incline to the central chamber. By now my eyes had adjusted to the gloom, and in a shaft of light up ahead I could make out the silhouettes of Neil, Josh and Ben.

Arriving at the core of the pyramid, known as the King's chamber, I took several gulps from my water bottle. 'Well done,' Neil said. Here at least there was space to stand and I looked around. Empty of antiquities save for a large stone sarcophagus with a missing lid, long ago destroyed by treasure seekers, I saw that there were relics here of a different kind. A small group of tourists dressed in hareem pants, tie-died t-shirts, beads and sandals were standing in a circle, making a sort of weird buzzing sound like a swarm of new-age bees.

'What are they doing, Mummy?' Freddie wanted to know.

I'd read that the pyramids attract all sorts, from those who hope to connect with and be taken by aliens, to others who think that 'pyramid power' (a kind of spiritual energy) can improve health, sharpen the mind and enhance sexual urges.[14] In normal circumstances, I viewed people like this as the worse kind of irritant, but as they were briefly taking my mind off thoughts of suffocation, it was quite fun to find them there.

'Oh, you know, they're just connecting with the place,' I answered vaguely. 'Some people believe that

---

[14] This kind of tourism is called metaphysical or sacred travel and is popular in countries with ancient civilizations like Egypt.

the pyramids hold a special kind of energy that can be harnessed.'

'Like a horse?' asked Freddie.

'More like a stream of energy flowing into their bodies. They plug into it and it charges them up.'

The boys all looked at me like I was talking rubbish.

'They're annoying,' said Ben. 'And I want to look at the tomb.'

'Well, we can,' I said taking a step towards the coffin as the hum intensified.

'Excuse me,' I said loudly. 'Can we get by?'

As the hum reached a crescendo, Josh was the first to spot that things were getting out of hand. 'They can't do that!' he shouted. 'Dad, make them stop!'

'What on earth ...' I gasped. A woman with long flowing grey hair had climbed into the sarcophagus. We watched as she raised her palms to the heavens and then squatted down.

'Oi, get out of there,' shouted Neil to no effect, as another of the group climbed in and then one more, until there were three in the coffin – all humming and squatting in poor King Khufu's grave.

A guard, who I surmised might well have been enjoying a sneaky post-lunch doze in a dimly lit corner, finally arrived on the scene and chaos ensued. He pulled at their arms in an attempt to oust them from the coffin. As one scrambled over the edge cloth was heard to rip. I hoped it was his stupid hareem pants with all that completely unnecessary material hanging from the crotch. While all this was going on, the other members of the group circled the tomb and continued with their incessant drone. Their behaviour was inexcusable, but the pure comic ridiculousness of the situation wasn't lost on Neil and me, and we began to giggle. The boys were horrified.

'Stop laughing,' Ben told us firmly.

'But it's funny,' I spluttered.

'It is not!' said Josh.

We pulled ourselves together.

'I think we should get going,' I said. By now the group were disassembling, and the thought of being trapped behind these utter nutters on our descent had brought on a fresh bout of panic.

As we hurried back through the tunnel towards daylight, the boys muttered to each other about what they'd witnessed. 'They should be banned from ever visiting the pyramids again,' suggested Josh.

'Banned from Egypt,' stated Ben, who, even at seven, had no time to spare for idiots.

Once outside, I leaned against the stone wall at the base of the pyramid and took deep breaths. Excited by all that they'd witnessed, the boys were keen to hang around to see what would happen once the group were turfed from the pyramid. One by one they emerged, blinking into the sunlight, like a festival of hippie moles.

'That one's not wearing shoes!'

'That one's got pink hair and is Grandma's age!'

'That's the one who was dancing in the coffin!' said Freddie, pointing to a woman with straggly grey hair wearing John Lennon style glasses with round yellow lenses.

Never mind the leather waistcoat worn over a bare chest, the rainbow bandana, the crocheted skirt. The worst crime this lot were guilty of was definitely against fashion. The youngest, I guessed, was in his mid-sixties.

'They're so old,' said Josh.

'Naughty old people,' said Freddie with no small degree of relish.

Ah, there it was: the fact that it was adults acting so disrespectfully and bizarrely was what had got the boys so riled. Not so much the act itself, I realized, but the ages of the perpetrators. For once, aged nine, seven and four, they were the good and responsible citizens, and they were enjoying every minute of it.

Over dinner later, the boys continued to discuss our strange encounter, now joining Neil and me in giggling at the absurdity of it all. We'd ordered a mixture of local dishes to share, which had been placed in the centre of the table, and feeling relaxed, I'd just taken my first gulp of wine.

'Children wouldn't ever act that crazy,' stated Ben 'We wouldn't behave like that, would we Freddie?'

'Mum,' said Josh, mid chew. 'I think I can taste nuts in my food.'

I shot out of my chair and raced round the table. 'Spit it out,' I yelled, loud enough to catch the attention of the entire restaurant.

Josh discharged his food on to the tablecloth and I scooped up the congealed mess and shoved it in my mouth.

'I can taste almond. Pretty sure it's almond. Perhaps with a bit of pistachio,' I yelped.

'Everyone is looking at us,' said Ben.

Josh, under Neil's instructions, was now busy swilling his mouth with water and spitting it into a glass, while a circle of alarmed looking waiters gazed on.

'All okay, Madam?' asked the bravest one.

'I told you that my son has a nut allergy,' I yelled. 'Go and find out what's in that dish.'

A few moments later the chef came running out of the kitchen to confirm that the chicken we'd ordered was actually a roulade, stuffed with nuts including almond, pistachio and pine nuts. 'I'm so terribly sorry,' he said. 'I thought the order was for chicken with nuts.'

Josh, although shaken, was fine. No need for an EpiPen on this occasion, but it was a timely reminder of just how careful we must be. When we'd all calmed down and dinner had resumed, Neil began to chuckle. 'You looked like a complete mad woman, scooping up Josh's chewed up food and shovelling it in your mouth like that.'

In retrospect, I'm not sure why I didn't just take an uneaten piece from Josh's plate. 'Well, I wasn't really thinking straight, was I?' I reasoned.

'Hmmmmmmmmmmmmmm,' hummed Fred, with a very mischievous glint in his eye.

Good old, Fred. You could always rely on him to keep it real. After all, most of us are only a few hums away from crazy.

7

Mauritius

A Lesson in Flexibility

Josh age ten, Ben eight and Freddie four

17 *October 2010,*
Mauritius

*'Shake yer bum bum. Shake yer bum bum,' sang Freddie, doing
an imitation of what Neil and I had (apparently) looked like
while dancing to Sega music.[15] He was making me look more
like Mr Blobby than the Beyoncé I had imagined. I blame it on
the rum cocktails we threw back at the Rhum and Fish Shack.
'I'll keep them coming, shall I?' asked the waiter. A question
that is always so difficult to refuse.*

*A torch-lit barbecue restaurant was set up on the beach,
near enough to the ocean to feel the occasional spray.
I attempted to persuade the boys to try some lobster instead of
piling their plates up with chicken. We'd paid a set price for the*

---

[15] Originating in Mauritius, Sega is one of the island's major music
genres. With its origins in the music of slaves, it is usually sung
in Creole.

meal, so it was a little frustrating to see that they weren't taking full advantage of the seafood on offer, but only Josh did as I suggested.

'That is delicious,' he told me. 'I'm having lobster all the time from now on.' I've yet to break it to him that he most definitely won't be.

The beat of Sega, played by a five-piece band, filled the night air, and we tucked into our meal while enjoying the feeling of the sand between our toes.

'Oh my God, what's Dad doing?' asked Josh.

Last we'd seen of him he'd been staggering up the beach in search of a loo, but now here he was dancing in front of the band all by himself. We watched him do a kind of limbo-like dance move, which caused other guests to cheer. There was no stopping him now and with his usual don't-give-a-hoot-what-anyone-thinks-of-me attitude, he followed this rather risky move up with an even dodgier shimmy.

'Daddy's drunk dancing,' hooted Ben. 'Ha ha. Look at Dad, Mum! He's dancing with the singer now.'

A rather lovely lady in a colourful ruffled dress was attempting to teach Neil a few traditional Sega steps, which involved shaking his booty and no small amount of hip thrusting. As an English man, I felt that he might have bitten off more than he could chew, but what he lacked in natural rhythm he more than made up for in enthusiasm. Freddie and Ben were delighted by their dad's exhibitionism, but I was surprised to note that Josh didn't look quite so sure, glancing nervously around and no doubt wondering what the *new friends he'd made at the pool that day would think? Was his dad a dancing legend or a total embarrassment? It wasn't yet clear.*

'Let's join in,' I suggested. 'Come on, we'll never see these people again and Dad's having such a nice time.'

I didn't have to ask Freddie twice. He was the first up, twirling in the sand, then collapsing into a heap when he got dizzy. Ben hopped up and down for a couple of minutes and then did the splits, and then Josh joined us to moonwalk over the sand. By now the beach was jumping, but where was Neil? No longer content with being a backing dancer, he had joined the band and was busy shaking a maraca.

〜〜〜

I was at that happy stage in my career when public relations companies sent me invitations to write about gorgeous beach resorts, most of which, incredible though it may seem, I declined. A beach holiday alone didn't give me much to write about. However, on one particularly drizzly day, when I had a bad head cold and was dreaming of sunshine, a press release on a lovely small resort in Mauritius landed in my inbox. Stories on hotels in luxury beach destinations were hard to sell. 'What's the angle?' every editor would ask. I scanned the hotel's website for ideas. It had a rum bar, an award-winning chef, beautiful villas, a fine-looking spa, but what

SHAPE OF A BOY

five-star resort didn't? And then flicking through the photo gallery, I caught sight of a peaceful-looking man sat in the lotus position. The resort employed a full-time yogi.[16] I'd practised yoga since my early twenties and always tried to make time to take a class or two while travelling, so with a cunning plan forming I pitched my idea for a feature on a family yoga holiday. Within a week I had four commissions but hadn't yet told Neil and the boys.

'Wow, that looks amazing,' said Neil when I showed him a picture of the hotel.

'Doesn't it? And they've invited us for a week.'

'A week? Let's do it.'

'There's one tiny catch.'

'Oh?'

'There's a bit of family yoga involved.'

'A bit?'

'Well, every day.'

'How much exactly?' asked Neil.

'Just an hour in the morning,' I lied. I'd actually agreed to two sessions a day.

'It's not really my thing, is it?'

'Well, you've never really tried it have you?'

'No. And there's a reason for that.' I watched Neil

---

[16] In 2010 this was quite unusual. Now every five-star resort has jumped on the 'wellness' bandwagon, employing legions of yogis, healers and the like. One I stayed at even had a shaman, known as a 'master of ecstasy'.

fold over and attempt to touch his toes. 'I just don't bend.'

'Well, if you'd like to come with me on this holiday you might just have to learn to be more flexible in all sorts of ways,' I said through tight lips. I was losing patience now. In my pitch I'd said that Neil and the boys would be excited to try a new hobby that the whole family could share. I'd made us sound wholesome. In other words, I'd fibbed.

'And have you met our children? I mean, come on,' continued Neil.

He had a point – Josh and Ben were all about rugby, and the rambunctious, dirty nature of the game suited them perfectly. Freddie, at four, was such a flibbertigibbet, forever chatty and on the go, that the idea of him sitting quietly in the lotus position, or standing still in tree pose, was utterly ridiculous. To persuade them to try something that would require calm and peace, I'd have to be sneaky.

'Hmmmmm. Well, you can think of it as a trade-off for sleeping in the bed that Elizabeth Hurley slept in last Christmas,' I told him. 'And leave the boys to me, okay?'

'Liz Hurley? Well, in that case ... Ommmmm-mmm,' said Neil.

Arriving in Mauritius, we by-passed the busy main tourist resort of Grand Bay and headed to where our hotel nestled on the edge of a coral sand cove in the wild south, where forest and fields of sugar cane meet the Indian Ocean.

'You've done well,' commented Neil as he took in the surroundings and our accommodation – a thatched roof two-bedroom villa that came with a complimentary family of Indian Mynah birds, cartoonlike with their vivid yellow beaks and feet. We watched the boys arguing over beds, which was how all holidays began.

'By the way, when are you going to break it to them that they're here to do yoga?' whispered Neil.

'When I give them these,' I said, pulling the yoga t-shirts I'd ordered online from out of the case. Ben's was blue with a yellow smiley face on the front that had 'namaste' written underneath; Freddie's was light blue and was printed with, 'yogi in training' on the front; Josh's was white and had a cartoon of a rabbit doing yoga, which read, 'Buddha bunny'. How could they not want to get involved when they had such cool gear?

'Boys I've got presents for you,' I shouted.

Josh eyed his t-shirt. The other two had pulled theirs on and were now charging around the villa doing kick-boxing jumps, making me worry that they'd confused yoga with some kind of martial art.

'I'm not wearing that,' he told me. 'And anyway, boys don't do yoga.'

Ben stopped in his tracks and glanced down at his t-shirt. This was a boy who wasn't swayed easily by the opinions of others, but there was no way that he was agreeing to do something that his older brother thought was 'girly'.

'I don't want to either,' he said, all too predictably.

'Er, well actually that's not true. Our teacher is a man and most master yogis are male, too.'

With three boys, I was quick to try and stamp out any gender stereotyping that occurred.

'Dad doesn't ever do it,' said Josh. 'Only you.'

I looked to Neil, who was lying on Liz Hurley's bed, and coughed. 'But I'm willing to give it a go,' he said (like we'd rehearsed).

I didn't yet want to break it to Josh that he didn't have a choice. I'd agreed to supply photos for the publications I was writing for and they knew I had three sons.

'And it's really good to try new things,' I persisted.

'Yeah, but not yoga,' he said, screwing up his face like I'd just told him he was going to spend the holiday learning algebra.

'Why not?'

'It's not a real sport.'

'Well, actually you're right. It's not a sport because it's not competitive. It's what's called a practice or discipline.'

'Will we be told off a lot then?' asked Ben.

Clearly, I wasn't making this sound fun.

'Listen, just give it a go this morning, okay?' I reasoned.

Josh sighed and pulled on his Buddha bunny t-shirt reluctantly.

Persuading Freddie to give yoga a go wasn't a problem at all. Keen to impress Dheeraj, our yoga instructor, as soon as he arrived for our first class Freddie struck a pose. 'Look I'm a warrior,' he told him, lunging forward and shooting his arms up in the air in a rather wobbly warrior one position. Away from the sceptical older two, I'd sparked his imagination by mentioning warriors and how they built their strength by standing in tricky poses for a long time. It had worked a treat.

'Wow, you know a lot already,' replied Dheeraj, with his gentle smile and calm demeanour, a little of which I was hoping would rub off on us all. Neil had been travelling for work a lot recently and was feeling burned out and I, tired from managing the boys by myself from Monday to Friday, had begun to let things slide a little with regards to routine and behaviour. Noise levels at home had risen, bedtimes had got later and there was a touch of anarchy creeping into mealtimes (usually once I'd had a glass of wine).

'My teddy can do it, too,' Freddie said. He'd placed three teddies at one end of the yoga mat and was now attempting to put his favourite, Snowflake, into the pose.

'Stop being silly, Freddie. Anyway, bet you can't do this!' Josh said, standing perfectly still on one leg.

'That's called tree,' Dheeraj told him.

'More like a weed,' commented Ben, who was busy trying to get his foot to touch his ear.

My three sons were discussing yoga positions. It was a conversation I never thought I'd hear.

I looked over to Neil, who was wearing the 'Real Men Do Yoga' t-shirt I'd bought him. He was glancing at the wall clock. Only five minutes into our lesson, and it was obvious that he was counting the minutes until he would be reunited with his sunbed.

'Okay, shall we start? Relax your buttocks and chant Om three times with devotion,' Dheeraj told us.

I heard a snort of a laugh from Ben and a giggle from Freddie.

'Dheeraj said buttocks,' Freddie chuckled.

He had and he'd also told us to relax them. I swallowed back a laugh, not daring to meet any of the eight eyes I had upon me.

'Ommmmmmmm,' I chanted.

'Ommmmmmmm,' the others chanted in a rather lack-lustre, self-conscious way.

Over the next hour, the boys morphed into positions such as bridge, plank, mountain and fish, while Neil tied himself in knots trying.

'Bloody hell, I can't get into that,' he said, watching the rest of us sit happily in the lotus position. It was true. Cross legged, his knees remained at shoulder height.

'That's not natural either,' he said as he watched us slide into pigeon by sticking one leg behind and folding the other in front, before lying stomach down.

'Ow! That hurts,' I heard him cry as Dheeraj gently applied a little pressure to his back in an attempt to ease him nearer to the floor.

I thought I'd made it clear that yoga was non-competitive, but Josh and Ben were soon contrasting their abilities in their usual gung-ho way. Josh showed off his malleable back by curving into a perfect cobra, while Ben found that his talents were in lying on his back perfectly still in corpse pose.

'You may not believe it, but this is actually one of the most difficult positions,' Dheeraj told us.

'See Josh, I'm the best at yoga,' said ventriloquist Ben, without even moving his lips.

We carried on like this for three days. After which there was a small revolt.

'I think the boys need a day off from yoga,' Neil said, meaning he did.

'Oh? But they're doing so well.'

'Yes. But a little break to do something different would be good. Did you know that you can swim with wild dolphins here?'

I had heard. It sounded terrifying to be honest – swimming out at sea, in water so deep that you couldn't see the ocean bed. No thank you. It wasn't the dolphins that worried me so much as what else might lurk below the surface.

'No. I didn't know that,' I lied. 'I'll find out about it.'

'No need,' Neil told me. 'It's all arranged for tomorrow. Thought I'd surprise you.' At times, when it came to getting what he wanted, Neil could be as sneaky as I was.

With a day off from yoga, we headed to Tamarin Bay, our departure point for the dolphin-swimming trip, driving over mountain roads where locals were out in droves picking guavas, the Mauritian equivalent to blackberry picking.

'I'm thirsty,' said Ben only half an hour into the trip. He'd caught sight of the stalls that sold sugar cane juice along the roadside, and with the sweetest tooth in the family, didn't want to miss out on trying it. We pulled up to one and read the sign, which claimed that among its many (dubious) health benefits, sugar cane juice could help with the growth of sexual organs, which predictably triggered a series of 'willy size' taunts. 'Your willy's so small that you need binoculars to see it', that kind of thing. Neil really did need to grow up.

By the time we reached Tamarin Bay to join the other ten or so (completely insane) tourists that had signed up to go on the wild swimming excursion, I was a bag of nerves.

'How far do we go out in the boat?' I asked our marine-biologist guide.

'Oh, quite a distance from shore. We're more likely to see dolphins the further out we go.'

I was having second, third and fourth thoughts about deep-sea swimming. I looked at my sons, who were all chattering at once and giggly over the chance of this adventure. Even Freddie, who at four was too young for wild swimming and would remain on the boat, was jumping up and down at the thought of seeing dolphins. They'd spent the last three days doing yoga, gamely trying everything that Dheeraj asked of them, so there was no way I could refuse to get on the boat. However, I could choose not to swim and with that reassuring thought I climbed aboard.

'You go ahead. I'll take photos,' I told Neil, trying to sound as casual as I could, once the skipper had cut the engine and we were adrift in open water.

'You're kidding?' Neil said. 'You're not really going to miss the chance to swim with wild dolphins?'

It did sound ridiculous put like that. It was like refusing tea with the queen, or flying to the moon or dinner with Brad Pitt.

'One of us needs to stay with Freddie anyway, and I'm fine just to watch from here,' I persisted.

Good old Fred, so handy to hide behind when I got scared.

'No problem,' said our guide. 'You can take turns with your husband.'

'Oh, but is there time? I don't want to be a nuisance and hold you up. It's really not . . .'

'We have lots of time,' he said.

'Also, I'm feeling just the tiniest bit sick at the moment and perhaps swimming will make it feel worse.'

'If you're feeling seasick, you'll feel much better off the boat and in the water.'

With an icy dread, it dawned on me that the hotel had most likely given them a head's up that I was a journalist and that, of course, they were hoping I would write about the experience, waxing lyrical about how wonderful it had been. But first they had to get me into the water. I caught an exchange of glances between skipper and guide, which told me that they no intention of letting me off the hook, and Neil had cottoned on, too.

'I think you're going in whether you like it or not,' he whispered. 'So, if I were you, I'd grow a couple.'

'Oh, shit. In that case, let's get it over with. Come on, Josh,' I said. 'Hand me those bloody flippers.'

'Don't you like dolphins, Mum?' asked Josh.

Not like a dolphin? That was like saying you didn't like Mother Teresa, or Gandhi or Father Christmas.

By now everyone on the boat was staring at me, asking the same question: exactly what kind of dolphin-hating nut job was I?

'Promise you'll stay with us,' I begged the guide, as I clung to the bow. They'd got me into the water, all they had to do now was peel me off the boat. My cover was blown, so there was no point in pretending any longer that I was anything less than terrified. I was not the intrepid travel writer they had been expecting, but rather a wimpy mum in a Boden bikini.

'I'll be by your side at all times,' he assured me. As usual, in these situations, I reminded myself that it wouldn't be good for business if a shark ate a journalist (although who would inform the shark of this, I wasn't sure).

Josh and I held hands tightly as we snorkelled along and the boat cruised away (hang on, no one mentioned that would happen). 'Where's it going?' I yelped, looking back to where Ben was now just a pinhead waving.

'It's so that the boat won't disturb our natural encounter with the dolphins,' the guide explained after I'd come to a halt (or as much of a halt that you can come to while treading water out to sea). Apparently, the skipper had heard by radio from another boat that a pod had been sighted and was headed our way.

'Keep your face in the water,' the guide told me. By now I rather suspected he may like me to keep it under permanently.

'Not happy, not happy at all,' I muttered through my snorkel, sending bubbles upwards and filling the tube with seawater.

Through shafts of sunlight, within a matter of minutes, around fifty dolphins appeared, and the high-pitched calls of their otherworldly chatter filled my ears. Although my heart continued to race, I was conscious of what an amazing moment this was, made all the more special by feeling Josh's hand in mine. They swam directly below us, barely 3 metres away, and with my free hand I pointed at each and every one, anxious that Josh wasn't missing a thing. When a mother with a baby swam by I discovered that it was hard to grin with a snorkel in your mouth. The positive energy and unlimited love that dolphins generate is proven (I might be making that bit up) and I felt it rise up through the water, like I was plugged into something universal, beyond my understanding (again, could be nonsense). Whatever it was I felt at that moment was nothing short of euphoria and left me gibbering like an idiot.

'Amazing, that was amazing, absolutely amazing!' I muttered, over and over again. I looked at Josh who was grinning from ear to ear.

'Are you crying?' asked Ben when I'd clambered back onto the boat. I hadn't realized, but I was.

'I think I am. Oh my God, I almost didn't go in,' I said. 'I could have missed seeing them.'

Ben gave me a hug. 'Well done, Mummy,' he said.

'You enjoyed it?' asked the guide.

'It was incredible,' I blithered.

As we cruised back to shore, I listened to Josh and Ben comparing notes on their dolphin encounters ('Mum was peeing her pants,' I heard Josh say), while Freddie sat up front with the skipper pretending to drive the boat. Perhaps it was time to admit that my adrenaline-junkie family were more cut out to adventure trips such as this, rather than yoga.

On the last morning of the holiday we were in our usual state of chaos. I was scrambling around trying to locate missing things – swimming goggles, flip flops, Top Trumps cards and toy cars that had been casually abandoned throughout the week, and Neil was without his passport.

'For goodness sake, where did you have it last?' I yelled at him.

'It's been in the safe the whole week,' he said, searching again. 'But it's not there now.'

'It can't have just disappeared.'

'Well, it has!'

The phone was ringing again. It was the third call to remind us that the airport taxi was waiting.

'Every time,' I shouted. 'We do this every bloody time! We've had a lovely week and now this stress!'

'Mum,' said Ben, who was standing in the doorway watching us.

'What is it?' I snapped.

'I think it might be a good idea if we all relaxed our buttocks.'

8

Sri Lanka
A Lesson in Karma

Josh age eleven, Ben nine and Freddie six

31 December 2011,
Uda Walawe National Park, Sri Lanka

Big drama in camp last night. Freddie had a nightmare and woke up screaming. He was so involved with his dream that it took me a few minutes to calm him down and by that time our tent was surrounded by armed guards. Neil, Josh and Ben, who were in the tent next door, also woke up and came rushing out to see what all the fuss was about. They'd obviously assumed the worst – that a leopard had managed to unzip the tent, or a tarantula had joined us in bed (the fringed ornamental tarantula, endemic in Sri Lanka, is one of the deadliest in the world, even though it sounds like it's dressed up to go to a party). I assured them that all was fine, but the guards insisted on scouring the tent for unwanted guests.

'We just check for naja najas,' *one said, shining a torch under the bed.*

'Naja najas?' *I asked, hoping above all that this wasn't some kind of snake.*

'Sri Lankan cobra,' he replied, confirming my worst fears. 'They like warm places.'

'Like safari tents?'

'And beds.'

'Do you often see them?'

The guard waggled his head in a non-comital way.

'Are they dangerous?' I asked trying to keep my voice calm and measured.

'Do not worry. We have excellent first aid,' he told me.

I'm sure this was meant to reassure, but for the remainder of the night Freddie and I huddled in one bed under blankets, both absolutely terrified. In daylight, the camp looked serene again, and we watched the sunrise over the river and kingfishers flit along the banks, while breakfasting on hoppers – a bowl-shaped pancake made from rice flour and coconut milk.

Out on safari we saw a herd of elephants flicking up red dust to cool their backs and pelicans perched on skeletal white trees, but Ben made the best and most surreal spot of all – a wild buffalo slumped in a mud slick with a tortoise sat on its back – like something from Aesop's Fables.[17] *When we arrived back to camp there was a fire of coconut husks lit and mugs of hot chocolate waiting for us. While I write this, the boys are*

---

[17] Aesop's Fables are believed to have been written in Greece in the late to mid-sixth century BC. In these moralistic stories, animals are often placed in human situations.

swinging in hammocks strung between trees and Neil is poking contentedly at the fire. What a happy way to end the year.

~

On the flight into Colombo, the capital of Sri Lanka, I'd begrudgingly filled in 'homemaker' (the best alternative to 'housewife' I could muster) in the occupation section of my landing card. Since the civil war, Sri Lanka remained a potentially dangerous country for journalists, ranking lowest with regard to press freedom of any democratic country. Even writing about something as frivolous as travel could be met with suspicion by the government, which controlled the majority of all media. So, to avoid any hassle I was keeping my job quiet. I realised that I'd forgotten to brief the boys, so leaned across the aisle to where each was playing on his Nintendo DS.

'Anyone want some sweets?' This was the only question that would get their attention when I was competing with anything on a screen. 'Don't tell anyone I'm a writer, okay?'

'Why?'

'Because the government here doesn't like foreign journalists.'

'Why?'

'Just because. It's complicated. Understood?'

They all nodded, anxious to get back to Mario Kart and I was relieved. It was becoming increasingly difficult to get away with an explanation of 'just because', but I didn't want to scare them with tales of abductions and worse. As plane journeys were the only times when they were allowed to stare at their screens for hours on end without Neil or me objecting, they were taking full advantage.

It was early morning in Sri Lanka when we touched down and we hit the ground running in an attempt to fight jet lag. In the run up to any trip, I'd engage the boys' interest by talking about the wildlife of the country we were visiting. In Sri Lanka, it was all about the Sri Lankan elephant, a subspecies of the Asian elephant, native to Sri Lanka and listed as endangered. After meeting Bonni, a short, stocky, instantly likable guy in his mid-sixties who would be our driver for the next ten days, we headed straight to Pinnawala Orphanage, home to around eighty abandoned or motherless elephants. The boys had seen working elephants in Thailand and Cambodia (sometimes badly mistreated), and I hoped this would provide a different perspective, with the elephants well cared for and treated with respect. They were tucking into breakfast when we arrived and a passing mahout (keeper) handed the boys a banana each to offer his young charges that were soon sniffing out the treats with their soft dextrous snouts. We watched enthralled as the trunks curled around the fruit and deposited them to their mouths. Lovely though the moment was, I was distracted. Further on, a larger bull elephant was chained, displaying the kind of distressed behaviour I'd seen other captive animals use, in his constant rocking backwards and forwards, and tossing of the head.

'Why's that one chained?' I asked Bonni,

'I'll ask,' he told me, turning to the mahout.

After a brief discussion, Bonni shrugged. 'All he says it that the bull is a danger to the others so they must chain him.'

'All the time?'

'I think so.'

Bonni must have registered my discomfort because he glanced at his watch. 'You could go down to the river to see the elephants bathe before we go,' he suggested. He was talking about the daily ritual when the mahouts drove the elephants from the orphanage to a nearby river for bathing. They're such beautiful beasts, but to be honest they make me nervous. I held my breath as they trooped sedately by, conscious that although they looked like gentle giants, fluttering their eyelashes coquettishly as they passed, they could easily wreak havoc at any given moment should they choose to. I was more relaxed once they were safely in the water, splashing one another with their trunks, and we could watch their antics from a distance. I noted that the elephant in chains wasn't getting a bath that day.

'I found that a bit stressful,' I admitted to Neil, who just looked at me blankly.

'The elephants so close to us. I wasn't expecting that,' I went on. 'Not sure there's been a proper health-and-safety check on parading them down here. Can't be safe, can it?'

'Kate,' Neil began patiently. 'You do know that there are thousands of wild elephants wandering around Sri Lanka.'

'Of course,' I lied. 'But it's unlikely that we'll see any. Bet they stick to the jungle areas.'

'No. It's common to see them in unexpected places. Actually, it's quite a problem. Farmers, in particular, often clash with them. They trample crops, destroy buildings and have even killed farmers who were trying to protect their land,' said Neil. Rather annoyingly, he'd, all of a sudden, become an expert on Sri Lankan elephant behaviour, and I'd been shown up for not having a clue.

'Alright, David Attenborough,' I muttered.

What you don't realize from looking at a map of Sri Lanka is just how long it takes to travel from A to B. Distances that look like they'd take about an hour, end up taking three, but the island is so densely populated that the view is never dull. 'Look at that,' shouted Ben constantly from the back seat of the minibus. He was pointing at a garishly hand-painted truck advertising the latest Bollywood film; at the piles of sour bananas and tiny sweet pineapples sold on the stalk at roadside stalls; at the little boy crying as his mum soaped up his hair next to a lake that would serve as his bath tub; at the Morris Minor cars, remnants of British rule; at the rice paddies dotted with storks standing stock still on one leg. There was just so much to see. Bonni chatted as he drove, telling us about the day the Tsunami hit in December of 2004.

'I was travelling inland on my way back home from Galle, after dropping off a Dutch couple on their honeymoon that I'd been driving for, when I heard the news. I turned the bus around straight away and drove back – my plan was to get them out of there, but I hadn't yet realized how bad it was. Many of the roads were closed and I couldn't get back to the coast to reach them. Instead, I picked up people from the roadside. One German man was trying to reach his family. Another Sri Lankan woman had become separated from her son. Then, I filled up my bus with those who needed to reach Colombo. I worked like this for two days, driving backwards and forwards.'

The boys were quiet, listening as Bonni described the scenes of utter devastation and human anguish he'd witnessed that day. Bonni had acted without a thought for his own safety, willing to risk his own life for the Dutch tourists he'd been looking after, and I could see that my young sons were both gripped and moved by his story.

'What happened to the Dutch people?' Josh asked.

'I don't know,' said Bonni.

'I like Bonni,' whispered Freddie. 'He's like a Sri Lankan Grandpa.' As compliments go, this was a very special one.

In Dambulla we planned to explore the Buddhist rock temples that date from the first century BC, visit Polonnaruwa, once the country's royal capital, and also climb Sigiriya rock fortress. Our base, the Amaya Lake Lodge, was idyllic with wooden chalets dotted around manicured gardens that led down to a lake. Armed with our bird-spotter's guide, Neil and I stalked through the gardens, spying kingfishers, Ceylon blue magpies and crested drongos, while the boys swung Tarzan-like on ropes strung from the gnarly old branches of kumbuk trees.

'It's lovely here,' I said. 'I don't want to visit anywhere else where elephants are held captive, even if it's in the name of conservation.'

'Those elephants were orphans though,' Neil reminded me. 'What would have happened to them without their mothers in the wild?'

He had a point and (annoyingly) I didn't know the answer, but still I couldn't shake the image of the chained adolescent bull. 'This is what I want the boys experience of Sri Lanka to be. Look at them over there, at one with nature.'

Occasionally I got like this, all misty-eyed and unrealistic, but fast-forward a few hours and it was a completely different story. All thoughts of an ideal world, where man and beast were at peace, gone. In the roof of our lodge, a wild animal had taken up residence.

'It's a cat,' the night porter told us.

'What like a domestic cat?'

'Huh?'

'Like a pet?'

'No. From the jungle. Too big and wild to keep as pet.'

'How big? Like this big?' I asked, gesturing with my hands the size of a small dog, let's say a dachshund.

'No bigger,' he said, opening his arms to the size of a Labrador.

'You're joking, right?'

'Yes. Ha! Ha! Very funny,' he said, indicating the dimensions of a West Highland terrier. Even so.

'Is it safe?'

'Yes. Very safe. He will be gone hunting soon.' Well, that was reassuring.

The (bloody) cat (which we never saw) prowled and scratched among the rafters all night, sounding like it might, at any moment, crash through the ceiling and join us in bed. It was obvious that it was taking a night off from hunting.

'Aren't you hungry?' I found myself shouting at around 3a.m., which served no purpose other than to wake everyone else up.

I barely slept a wink and was relieved to see the first flickers of dawn and hear the alarm go off at 6a.m. for our morning climb of Sigiriya – an immense chunk of hardened magma, once the plug of a volcano, long since eroded away. Shaped like a parcel with steep metal steps bolted to the sides, it would be a challenge to climb it even with a good night's sleep. Said to be the site of King Kashyapa's palace in the fifth century, later archaeological finds suggest that perhaps Sigiriya was first a Buddhist Monastery, existing since the third century BC. The king, they say, was carried to the top by nimble-footed servants, and after scaling the first couple of hundred of these precipitous steps I was beginning to understand why.

'You okay?' I asked Freddie.

'Yep,' he told me. I'd assumed that for Freddie's six-year-old legs the climb would be quite a struggle, but I was wrong. He had something to prove in not letting his older brothers beat him to the summit and kept up the pace just fine.

'Fit boy,' commented one saffron-robed monk as Freddie overtook him.

Higher and higher we climbed, staring down on the roofs of the treehouse watch towers that night security guards slept in, safe from the wild elephants who liked to roam here after dark. It was yet another reminder of how Sri-Lankans lived cheek to jowl with their wildlife. Halfway up, we came to the Cobra Hood Cave, named for its overhang that was shaped like a snake's head. In the cave, a snake charmer had taken up residence (of course) and informed us that his snake would 'dance' for one hundred rupees. Egged on by the boys, Neil handed over the cash and we watched the snake uncoil and rear its sleepy head before nestling back down in its basket.

'Is that it?' demanded Josh.

'What did you think it would do, break dance?' I asked.

'Ha! You've been ripped off, Dad,' Josh said, as if he'd warned Neil all along that might happen.

More impressive were the views at the top of Sigiriya and we gazed over a wide sweep of jungle, speechless both from the beauty of it all and from being out of breath. We picked our way through sparse ruins until we found a stone slab – either the base of a king's throne or a monk's meditation spot. With no information signs (a huge disappointment for Neil) I rather liked the way you could make it up as you went along.

'Not much up here,' commented Freddie. I'm not sure what he was expecting, perhaps a small café selling ice cream or a gift shop. My sons' often unrealistic expectations never failed to surprise me. 'And I need a wee.' Of course, he did. Someone always did (although usually it was me) in the most unsuitable places. Unluckily for my sons, they were not in possession of a Lady-pee, which I kept hidden deep in the bottom of my day pack.

'You'll have to hold it,' I told him.

'I can't,' he said.

I spotted a warden taking a tea break under a tree and bounded over. 'My son needs to go to the toilet,' I told him.

We were very enthusiastically led to the edge of the site, where a small ruined building stood.

'Please use the King's,' he told me, holding out his hand for a tip. Inside, we found a stone seat with a hole hollowed out. Whether it was King Kashyapa's loo or not, I'll never know, but Freddie was chuffed with the idea of taking a royal pee.

In heat rising to thirty degrees by late morning, we were glad to be back in the minibus and heading to The Amaya for a few lazy hours of swimming and napping (cat free), when Bonni suddenly screeched on the breaks and we were thrust forward in the bus.

I'd been looking out of the back window, trying to take a shot of a train that was passing behind us, packed with commuters, many of whom (without tickets) hung from the doors, ready to jump if an inspector appeared. I hadn't seen what was ahead of us, but the others had.

'Bloody hell,' I heard Neil shout.

'Whoaaaaaa,' the boys hollered in unison.

'Stay calm,' I heard Bonni say, as the boys all scrambled to the front of the minibus for a closer look. 'He's not interested in us.'

I pushed through them all to see what all the fuss what about to discover that 'he' was a large bull elephant that had emerged from the bushes on the side of the road. 'He' was only about 20 metres away from our minibus. 'He' looked angry, tossing his trunk around and swaggering backwards and forwards in front of a tuk-tuk, whose driver was scrambling out of his seat and running away, losing a sandal as he scarpered.

'Shit,' I shouted. 'Shouldn't we get away from here?'

'Shit,' echoed Freddie.

'Shit,' said Ben.

'Shit,' shouted Josh.

We had a new (stupid) rule in our family that if Neil or I swore, the boys were allowed to repeat, just once, what we'd said, which they were taking full advantage of.

After being shown up for a lack of elephant knowledge at Pinnawala, I'd done some reading and Neil had been right: in a country just shy of twenty-two million people, an estimated seven thousand five hundred elephants roamed free, and the relationship between man and elephant wasn't an easy one. Now fully aware of how dangerous an angry bull could be, I really couldn't understand why Bonni was so calm.

'He has no argument with us,' he said. 'He's been waiting in the bushes over there for that tuk-tuk driver to pass by. The man must have mistreated him at some point.'

'Are you sure? I don't know about that. He could charge down here at any moment,' I said, glancing to the gridlocked traffic behind before saying a worse word than 'shit', which the boys didn't dare repeat.

'An elephant will not harm you, if you have not harmed it,' repeated Bonni.

'Yeah, but is that really true?' I persisted, much less composed than the four males in my family, who were all eagerly glued to the action.

Bonni had told us a story earlier in the day about an elephant in the village he grew up in that always returned to a pineapple seller at exactly the same time each week, who would treat the elephant to one of his juiciest fruits. That was a cute story, but this?

'Treat an elephant with disrespect and he'll always remember,' said Bonni, as we watched the elephant pace around like he was deciding what to do. Decision made, he took a few steps backwards, then charged at the tuk-tuk and flipped it over with his trunk and tusks. It lay in the road, wheels spinning, like a toy car. The elephant gave it a cursory kick and swaggered back the way it had come, crashing through the bushes.

'That was so cool,' said Ben.

'Karma,' whispered Bonni. 'That is what it was.'

Like the majority of Sri Lankans, Bonni was Buddhist and so his belief in Karma – the cycle of cause and effect, that a person's good or bad actions will dictate their future – made it easy for him to explain the elephant's behaviour. Buddhists also believe in all living things having equal respect. It felt like too good an opportunity to miss as a parent. This was a big one when it came to life lessons, and whether you called it treating others how you wish to be treated, or simply what goes around comes around, I was keen for my three sons to take something away from this dramatic event.

'So, that's why you need to think about your actions and always behave towards others how you would like others to behave towards you. Right, boys?'

'There aren't any elephants in Windsor,' Freddie rightly pointed out.

'But think about the bigger picture. How you treat friends, your pets, people you meet as you are going about your day, your brothers...'

I lingered over the word 'brothers', hoping the boys would get the message. Recently, there'd been a spate of quarrels between Josh and Ben, which had led to punch-ups.

'And Daddy's car would be harder than a tuk-tuk to flip over,' added Freddie, missing my point entirely.

'What I mean is, always think about how you're treating each other. Good Karma produces more good Karma, isn't that right, Bonni?' It was always useful to get another adult on board that perhaps the boys would take a little more seriously than Neil or me.

Bonni nodded. 'That is right. The elephant has been there for a while. I noticed it when I drove past here last week and now I know what it was waiting for. The tuk-tuk driver had no idea, had most likely forgotten about the elephant he mistreated. We won't see the elephant again, now that he's taught him a lesson.'

The traffic began to crawl forward. The tuk-tuk driver was back and had retrieved his sandal, which he was waving angrily in the air. Other men had run over and were attempting to roll his tuk-tuk upright by rocking it backwards and forwards, but the vehicle didn't look in good shape. My heart was still pounding but the boys were only flushed from the excitement of it all.

'Serves you right,' shouted Ben out of the window as we motored past, as if the tuk-tuk driver's day wasn't quite bad enough already.

'Karma,' said Freddie, trying out this new word. 'Right, Mum?'

'That's it, Freddie. Karma,' I repeated, as my heartbeat slowed to normal. What we'd witnessed had brought a whole new perspective on that cutesy little maxim that 'an elephant never forgets'. What next, I wondered, an actual 'elephant in the room'?

9

Malaysia

A Lesson in Conservation

Josh age twelve, Ben eleven and Freddie seven

14 *April* 2013,
Langkawi, Malaysia

Oh, the utter joy of having a husband who is like Mr Bean. Today, Neil excelled himself in both the clumsy and dangerous stakes, during a ride on one of the world's steepest cable-cars. We made it to the top of Mount Machinchang without a hitch, where a 125-metre sky bridge is suspended 660 metres above ground level between a gulf in the rainforest. With a fear of heights, Neil couldn't look down, but with the boys help I coaxed him on. Further along, a Japanese lady was crying, frozen in fear; behind her, another tourist was on his hands and knees crawling back towards us with his eyes shut, muttering encouragement to himself.

'Ignore them,' I told him. 'You'll feel better once we get going.' I unpeeled his hand from the rail, determined that we would all make it to the lookout point at the end of the bridge with its

views to Southern Thailand. The boys ran on, all daring each other to jump up and down to see if they could make the bridge shake, causing the Japanese lady to scream.

'I'm glad that's over,' Neil admitted once we'd made it across and back and were waiting at the cable car station for our descent. High on the relief of being off the sky bridge, he was now cracking jokes and filming the boys as we boarded the car. Poor Neil, he must have thought he was almost home and dry until the doors slid shut and trapped his foot.

'Oh, my God. My foot! My foot! I've trapped my foot!' he shouted.

'Just give it a good pull,' I told him.

'It won't budge. Oh, my god. I'm stuck! There must be an alarm! Ring it now! Quick! Before we . . .'

At that moment the cable car swung out from the station and our descent began. Surely raising an alarm would just cause a lot of unnecessary fuss. Luckily Neil couldn't reach it.

'Try to calm down. You're frightening the boys,' I told him (to be honest though the boys loved nothing more than a bit of drama).

'I'm f**king stuck in the door of a cable car at seven hundred feet. Don't tell me to calm down.'

Neil was still filming and on realizing this, I began to laugh. 'And what the f**k are you laughing at?' he wanted to know.

Freddie knew that the 'F' word was saved for when things were really f**king bad, and began to look alarmed. 'Don't fall out, Daddy!' he cried, while sensibly edging away from Neil and the gaping door. The other two, however, looked thrilled by

the commotion. 'Mum! Mum! What are you going to do?'
they shouted.

I took the video camera out of Neil's hands. 'Oh, I'm going to
carry on capturing this epic family moment,' I told them,
pointing the camera at Neil. We reached the ground with Neil
caught on camera sticking two fingers up at me.

~~

I was so excited to be returning to the Malay island
of Pulau Langkawi (the largest of ninety-nine
islands in the Langkawi archipelago), which Neil
and I had visited during our travels around South-
east Asia in 1994. Sandwiched between the Thai
island of Phuket to the north and Malaysia's Penang
to the south, it is fringed with wide sandy beaches
with an interior of pristine jungle and paddy fields,
and tourism had boomed in the years since our
first visit.

'It doesn't look the same!' I wailed as we landed at
the airport.

'Well, there wasn't an airport here in 1994,' Neil
reminded me. That was true. Back then, we had
arrived by boat.

'Did you see all those hotels? I bet our lovely little
beach shack isn't there anymore.'

Neil raised his eyes. He was used to these totally
unrealistic expectations I had of places remaining

as they were in my memory. 'Nope! It definitely won't be,' he said.

'Was it a mistake to come back?'

'Bit late now,' he said, reaching for the boys' day packs from the overhead locker.

I thought fondly back to stepping off the ferry from Penang eighteen years before and of our journey by taxi through a sleepy jungle paradise to the island's most beautiful beach, Pantai Kok. Our budget only stretched to a basic chalet (about the size of a bathing hut), which made up for its lack of mod cons with a hole in the roof through which I could spy Sirius, the brightest star in the sky. Life was idyllically simple: we ate banana pancakes for breakfast, spent lazy hours watching the white-bellied sea eagles plunge in and out of the off-shore islands, and explored the island by rented motorbike. Like a backpacker's badge of honour I still bore the scar from burning my leg on the exhaust.

'And anyway, it would be hell on earth stuck in that tiny beach shack with those three,' Neil added, nodding towards Josh, Ben and Freddie, who were shoving each other around in the aisle. This time we weren't only lugging backpacks, but three spirited sons aged twelve, eleven and seven. Although, on the luggage score, I prided myself on keeping it light, often amazing check-in staff with our meagre belongings, namely three large canvas bags plus our individual day packs. Neil had a point though; we'd never squeeze into a beach shack.

Our home for the next week was a luxurious resort with private beach at Tanjung Aru, and as I checked in, I felt a little like I'd sold my soul. Neil and the boys, however, were blown away.

'Wow, we even have our own monkeys,' said Josh, looking up to where a troop of spectacled langur monkeys were feeding noisily in the cashew trees outside our family villa. The porter quickly advised us to keep the door shut at all times as the macaques that also roamed the resort had cottoned on that there were nuts in the minibar. In the outside rainforest shower we discovered a resident lizard, around a foot long. 'Don't worry. He's harmless,' we were told, so the boys named him Geoff. The loo had floor-to-ceiling windows, both handy for spotting those peanut-stealing macaques and for watching the tiny twitchy-nosed shrews that hid amid the traveller's palms and rattan in our private jungle garden. I had to admit it was a fine example of how a resort could blend successfully with its environment, and yet I was still uneasy about the impact of tourism on the island.

On day one I indulged the boys and Neil with a lazy decadent day of lounging poolside, and then on the second I made an announcement. 'So, about today. I've something exciting planned.'

'Beach volleyball?' asked Josh.

'Kayaking?' suggested Ben.

'We're doing what?' came the incredulous cry from my three sons when I explained that we would be spending a morning of our holiday picking up litter.

'It'll be fun,' I told them, smiling brightly. 'We're going to be eco-warriors. Doesn't that sound cool?' They didn't look convinced.

'We're going by boat into the mangrove with the resort's naturalist,' I continued. Always the involvement of another adult made it a good bet that they'd do as they were told.

'Not to be confused with naturist,' said Neil. 'So, remember to put on your swimmers. There'll be no skinny dipping on my watch.'

As usual the boys ignored Neil's 'dad joke' and continued to grumble.

'Exactly how long will we be out for?' Josh wanted to know.

'Why us?' asked Ben, looking miserably back to where another family was playing beach boules.

'Can I go to the kids' club instead?' asked Freddie, who never ever asked to go to a kids' club.

'You're all coming. End of,' I told them.

I'd snuck off the day before on the pretence of attending a yoga class, but instead had done some research. Of all the large luxury resorts on the island,

I was thrilled to learn that ours appeared to be making the most effort to involve guests in its pledge to help preserve the island's mangrove habitat and teach something about the ecology of the island. Since our last visit, Langkawi had been given an UNESCO Geopark status in order to preserve the maze of mangrove forest and limestone karst hills, some of the oldest in South-east Asia. Back in 1994, no one had cottoned on to how much the mangrove forests would be of interest to foreign visitors and there were no tours of the area. Now there was a steady stream of tourists armed with zoom lenses and binoculars zigzagging the waterways, many of whom often left more than footprints behind. Our resort was ideally situated for exploration with the main inlet to the mangrove forest just a few minutes away by boat. It wasn't like it even required much effort for us to do our bit for conservation.

Awang, the resort's full-time naturalist (pristine in Khaki, with a comb-over as low as a falling tide) was waiting on the beach with his boat, ready to take us out on our mangrove spring clean. The boys had accepted the situation by now, already quarrelling about who'd have the first go with Awang's 80-centimetre reacher-grabber tool, designed for getting to rubbish tangled in the mangrove that was beyond arm's reach. I gave it around thirty seconds before someone's bum was pinched and there'd be tears.

'Don't move around the boat and please be careful while leaning over. Although they're not commonly sighted, crocodiles do live in these waters,' Awang told them.

The boys laughed. 'Yeah, right,' I heard Josh mutter.

'I don't think Awang's joking, are you?' I asked, unsure myself.

'Never about crocodiles,' he told us.

In fact, as it turned out, never about anything much at all. Awang took his job very seriously. His rapport with children began and ended with the word 'don't', as in 'don't rock the boat, 'don't touch that', 'don't lean over' . . .

After an hour, we'd collected a bag of rubbish each, filled with petrol containers, plastic bags and water bottles either dropped or cast overboard from tourist or fishing boats, and a surprising number of flip flops. But there hadn't been a lot of laughs.

'There must be a lot of people walking around with one flip flop,' observed Ben. A strange game of 'pairs' was quickly invented, one that required you to find a set of flip flops, and the rule was that you could match by either size or colour. Their ability to find fun in every situation was a constant source of joy to me.

The novelty of the reacher-grabber had by now worn off, and Josh was trying to persuade Neil to dangle Freddie like a human fishing line instead.

A request that I suspected was designed to both scare Freddie and wind Awang up. Although on-board entertainment was a little lacking, offshore we were amused by a whole circus of unusual creatures, like the male fiddler crab with its one giant red claw, which it used to attract vivid blue females and to box rivals with. When the crab looked as if it might lose a fight, it cut a door in the mud to retreat into and slammed it shut. It was like watching a cartoon.

'There's a walking fish!' shouted Freddie, pointing to where the mudskippers shimmied over the glistening sludge, the males fanning out their top fins like sails on a ship.

We navigated the channels peering into the tangle of black roots to search for debris. The gentle langur monkeys gave us curious, shy glances from the treetops, while the macaques screeched shamelessly for attention. Pausing at a rock face, Awang pointed up. 'Take a look, boys,' he said. 'How much do you think that flower is worth?'

A diminutive flower stuck out from the rock face. 'Two pounds, fifty?' ventured Freddie.

'Quite a bit more,' Awang said. 'Take another guess.'

'A hundred pounds,' guessed Josh.

'Even more.'

'More than a hundred?' questioned Ben.

Awang nodded. He had their full attention now.

'It's a very rare, endangered orchid that I'm not allowed to even name to foreign visitors. On the black market, it can sell for three thousand pounds.'

'Let's pick it!' suggested Freddie.

Awang narrowed his eyes. 'Oh, you might want to reconsider that idea. For a start there's a big fine and you might even go to prison if you were caught.'

'For picking a flower?' asked Josh.

'Not just a flower. An orchid on the edge of extinction that's intrinsic to this island, part of a fragile eco-system that we need to protect. A flower so sensitive to the changes in the island's biodiversity that it acts as a general indicator for the health of this entire fragile mangrove habitat. If the orchid is blooming that gives us hope. If we lost this beautiful flower not only Langkawi but the world would be a much poorer place.' The boys stared at him like he was from another planet, but I think they got the message.

'Mum's always die,' said Ben.

Thanks Ben. Okay, so it was true that the few times I'd been given an orchid in a pot it hadn't gone well, but there was no need to tell on me to Awang. Ben was making me sound like some kind of mad-axe orchid killer.

'Not particularly green-fingered,' I muttered by way of explanation, while giving Ben a sharp poke in the ribs.

We sailed on into the gaping mouth of Crocodile Cave, which Awang assured me was named for its shape, rather than being a hangout for large predatory reptiles, and as our eyes adjusted to the gloom we spotted bats roosting, their eyes glowing eerily yellow. Cruising out the other side, Awang was on the look-out for fallen mangrove saplings, which he showed the boys how to replant by spearing them back into the muddy banks. Now he was winning them over. Anything competitive that involved lobbing something javelin-style while balancing in a boat was bound to be a hit. 'Don't throw them quite so far,' he said, just as Josh hurled one a good five metres, achieving the perfect plant.

'That okay, Awang?' he asked innocently.

Awang was also keeping an eye out for unlicensed boats, reaching for his camera to take zoom shots of the pirate skippers. The boys soon got in the swing of things – pointing out the vessels that they thought were driving too fast, and those that were churning out fumes or missing license plates, until I felt like an extra in an episode of *Langkawi Five-O*.[18]

As we disembarked from Awang's boat he invited us to join him on something called a jalan-jalan

---

[18] *Hawaii Five-O* is an American television police drama, which first ran from 1968–1980 and was later remade by CBS and ran from 2010–2020.

walkabout at 6a.m. the following morning. My family all looked at the ground.

'You'll learn something of the flora and fauna around the resort,' Awang explained. 'It'll be very educational.' Ah, he'd really blown it now by using the word 'educational'. This was a word I avoided at all cost.

No one said a thing. I fumbled for an excuse. 'Sounds lovely, but ...'

'And at that early hour we will have the resort to ourselves. Not many guests are tempted to rise at that time.'

No shit, I thought, as I watched Josh circle his toe in the sand.

'I'll come,' I blurted out at last, taking one for the team. 'Anyone else?'

'Kind offer, but someone should stay with the boys,' Neil said, finally lifting his head and grinning at me. 'You go, Kate. I don't mind. Honestly.'

I'd been played.

Next morning, I traipsed around the resort with Awang. Another guest had risen to join us – a retired English gentleman, who was hairy in all the wrong places, dressed in pastel shorts (horribly incompatible with his knobbly white legs) and wearing socks with sandals. Ridiculously chipper for the hour, he had a whole notebook of (bloody ridiculous) questions prepared for Awang. Childhood memories flooded back of being frog-marched round the Lake District by my Uncle Ron (a boy scout trapped in the body of an adult nuclear physicist) and having various tree names barked at me while he read from the *Oxford Book of Trees*. However, there was no way I was going to let Neil and the boys know how much I'd suffered. Keen to prove that it had been worth the early start, over breakfast I showed off. 'Actually, you missed quite a lot of interesting things,' I told them. 'For example, I now know how to make black boot polish from the hibiscus flower.'

I received not one glimmer of interest from my family, who were all busy tucking into pancakes.

'And I also know how to make alcohol from the nectar of the coconut flower.'

'Well done, Mum. If we ever get lost in the jungle in our school shoes at least we'll be able to keep them clean,' said Josh.

I'd accepted my family's need for a lie in, but that hadn't stopped me hatching another plan. Sensitive to our sons' general apathy towards nature excursions (aka adventure trails, wildlife quests, explorer's escapes), I went to great lengths to disguise what we were actually up to. I had booked a taxi to pick us up late morning to take us to a jetty where a ranger would be waiting to take us out on a Secret Voyage (God, I was good). In the 1990s the sea eagle population was declining so the government introduced a feeding programme that was managed by Geopark rangers and we were going to help.

'Good news that the eagles are thriving again, isn't it?' I said, as we motored out to the centre of a small inlet. The boys weren't talking to me as I'd wrenched them away from a game of manhunt at the pool with new friends, who had, and I quote, 'normal parents'.

We could see the eagles perched high up on the trees waiting for their dinner. From that distance and with their wings tucked in the birds looked quite small, but I knew once the eagles took flight, that the boys would have a shock.

'Ready?' shouted the ranger, casting what looked like a chicken's arse into the water.

One eagle caught sight of it and swooped in for lunch. With a wingspan of up 2.2 metres and a fearsome hooked beak, they were an alarming sight when flying straight towards us and Freddie covered his head as we watched it plunge with outstretched talons to retrieve the chicken from the surf.

'Whoa, it's massive,' shouted Ben. Now all three were glued to the action, making bets on which eagle would make the next successful catch.

'Want to throw some?' asked the ranger.

The boys peered into the bucket filled with blood, entrails and chicken carcass. 'No thank you,' they answered politely.

'See, that was fun, wasn't it?' I crowed as we clambered back into the taxi. 'Just one more stop and then back to the hotel, okay?'

I hadn't quite let go of the idea of rekindling a moment from our first trip to the island, so we drove across the island to check out Pantai Kok.

'Oh, I can't bear it,' I cried. 'It's nothing like I remember.'

'Told you this would be a bad idea,' Neil said.

'This is where our beach hut stood,' I told the boys, gazing at an empty spot on the beach. A string of new hotels, shops and restaurants had been built behind us, where I remembered forest. The eagles that I'd watched on their way to fish and then return to nest were nowhere to be seen.

'Don't be sad, Mum,' said Freddie, but it was difficult not to be.

'The eagles have gone,' I said.

'Yeah, they're all gobbling up chicken dinners on the other side of the island these days,' Neil said. 'They're not stupid.'

They humoured me for five more minutes while I continued to mourn the passing of Langkawi in 1994, before Neil suggested we stroll along to the swanky new marina for an ice cream.

'Well, it's done now,' he told me, when I shot him an exasperated look.

Back at the resort, the boys played football and

volleyball on the beach; and had their lunch at the pool bar, drinking mocktails and eating pizza while sitting waist deep in water. I left it two whole days before I even mentioned that we were going on a 'fowl prowl' (quite possibly my best deception yet), which was actually a hike up Gunung Raya, the island's highest mountain, to look for hornbills and toucans, two of Langkawi's most varied, large and noisy birds. I'd been told by Awang we'd see plenty.

'Where are they then?' asked Josh, losing patience after only half an hour, when the game of trying to get the word toucan into a sentence, as in 'toucan play at that game', 'one person can't play tennis, but toucan', and the best from Ben, 'it's toucan fusing' had run its course. Another hour passed, and about to give up, Neil looked up and spotted something sitting in the top branches of a soaring mahogany tree.

'Looks like a hornbill,' I said, peering through binoculars. I quickly thumbed through the Birds of Malaysia guidebook I'd borrowed from Awang. 'Bloody hell, I think it's a wreathed hornbill.'

If I was right, this was one of Malaysia's most elusive birds. I quickly changed lenses on the camera for a longer zoom

'If I could just get a little closer, I'd be able to get a great shot of it,' I said, hanging over the side of the mountain path.

'I reckon I could climb up that tree,' said Neil.

'Mr Bean goes bird-watching,' said Josh, as a swift reminder to his dad of quite what a bad idea that would be.

As if to settle the matter, the wreathed hornbill flew off out of view. 'Pool time?' asked Freddie hopefully.

To salve my conscience, the boys had played their part – gamely picking up litter and taking part in nature excursions without too much of a moan. I had to concede that it was now time for footy on the beach and games of manhunt in the pool. I might not have got a photo of the hornbill, but even a glimpse of this secretive bird had been a windfall.

On our final evening we sat on the beach and gazed out to the neon glow of the fishing boats that were out to lure squid to their nets. We'd just enjoyed a barbeque of Langkawi green Spiny lobster, chicken kebabs and butter fish and were now happily toasting marshmallows over the fire for dessert.

'I've decided what I want to be when I grow up,' Josh told us.

'Oh, yeah?' I was only half listening to be honest, caught up in the scene out to sea, and expected the usual answer of pro skateboarder.

'A marine biologist.'

I sat bolt upright. Had my twelve-year-old son really been so inspired by all he'd seen in Langkawi that he was now prepared to dedicate his life to marine conservation? It was a mother's dream come true.

'Wow! Really?'

'Yeah, those fiddler crabs really crack me up.'

Ah, okay. That was more like it.

10

Mexico
A Lesson in Mortality

Josh age thirteen, Ben twelve and Freddie eight

19 April 2014,
Mérida, Yucatan, Mexico

Great choice of restaurant tonight and one the boys will remember thanks to the abundance of a certain 'disgusting bog weed' (Ben's description).

'What the hell is that?' Josh asked. It was in fact my Chaya margarita, and looked like something Shrek might bathe in – bright green with a swamp-like consistency. The boys watched with utter disgust as I attempted to pull a particularly stringy piece from between my teeth. Chaya is a Mexican superfood, more nutritious than any other leafy green, and used in everything from scrambled eggs to ice cream to cocktails. When our tacos with pulled pork arrived, they were also sprinkled with the stuff. Freddie took a nibble, while Josh and Ben got to work straight away in scraping it off.

'Oh, for goodness' sake, just try it,' I said.

'This actually makes me cross. Why would anyone want to ruin such nice food with this?' Josh demanded.

'You can eat mine,' Ben said, pushing his rejected Chaya on to my plate.

To be honest, I've had my fill of Chaya and would have liked to have had a word with the bright spark who'd come up with the idea of ruining a good margarita with it.

After dinner, we headed back to our hotel for 'the show'. Inspired by all the mariachi bands we've seen around town, Freddie bought a small wooden guitar in the craft market in the central plaza today. Settling into our hammocks, we watched him pluck at the strings.

'What are you going to play?' asked Neil.

'This,' said Freddie. 'Don't you want me, baby? Don't you want me, ahhhhhhhh. I was working as a waitress in a cocktail bar . . .'

How Freddie knows the lyrics to this Eighties Human League hit is anyone's guess. His spirited rendition of 'Don't You Want Me Baby?' might not have been what we were expecting but was comedy gold.

'Know any others?' I asked hopefully. 'That was brilliant.'

'Jingle bells?' he offered.

Carefree, uninhibited and full of joy, Freddie's performance is so in keeping with the city of Mérida. From our balcony, we watched as elderly couples gathered to dance in the plaza below. The women wore slinky silk dresses and flowers in their hair; the men were in crisp white shirts and braces, and they *swayed to the music, perhaps lost in memories of being in their twenties, or more likely content to be the age they were now.*

'That man must be as old as Grandpa,' commented Ben, as we watched a silver-haired Fred Astaire twirl his senior-citizen partner.

'Probably all the Chaya they eat,' I said. 'Just saying.'

'Nice try, Mum,' said Josh. 'But we're still not eating it.'

~

Mexico was to be a care-free trip; a country to raise our spirits, where we would drink tequila, explore Mayan ruins, snorkel with turtles and lose ourselves after a very difficult year. My sister's lovely husband, the boys' much-loved uncle, had died unexpectedly of lung cancer and I knew that losing Rob had affected our sons more than they were letting on. The realization that death could come so quickly and mercilessly was something that they were still grappling with, and I was relying on Mexico, with its colourful culture and sun-filled days, to give us the respite we all so desperately needed.

Let's be honest, our Western culture doesn't deal with death well. Even our plan to take the boys to Rob's funeral proved controversial, and all, without exception, thought Freddie was too young to attend. Swayed by other's opinions, Neil and I discussed the option of leaving the boys at home, but it was never really a serious consideration. I felt they needed the opportunity to say goodbye, and as Rob was a Buddhist I knew that his funeral would be an uplifting occasion with chanting to help him on his way to a new life. Even so, it was impossible not to be sad. Months after, Freddie had remained a little anxious, worrying about what would happen if Neil or I died.

'You okay, Mummy?' he'd ask peering at me.

'I'm fine.'

'Are you sure?'

'Yes, absolutely.'

'Do you have a headache?'

I did, and I'd been stupid enough to mention it. 'Not a bad one,' I assured him.

On a daily basis Freddie gave me a health check but providing reassuring answers to my eight-year-old often proved difficult. The answer of 'Nothing's going to happen to me,' was met with the very astute, 'But how do you know?' Freddie was of an age, when he couldn't be fobbed off with a cuddle and a promise that I would always be there for him. He was right. I didn't know that for sure.

A holiday in the sun was just what the doctor ordered. On arrival in Yucatan, we bypassed busy Cancun (foam parties weren't on this prescription) and headed instead to Hacienda Chichen Itza, built in the sixteenth century by Spanish conquistadors, and located in the grounds of Chichen Itza, Yucatan's most celebrated Mayan ruins.

'Is this really where we're staying?' Josh asked, exploring our cottage where archaeologist Dr Sylvanus Morley, head of the first Mayan expedition, stayed while reconstructing Chichen Itza's temples in the early 1900s.

'It's like a museum,' said Ben, sitting at the desk where Morley might well have sat to pore over excavation plans.

It was true that little had changed since that time (including the plumbing judging by the noise the toilet made when flushed), and it was simply decorated with hand-carved dark wood furniture, with the kind of bed you might expect to find the big bad wolf lying in.

'I like it,' I said. 'It's got character.' Character was the rather vague description I used when somewhere was historic but in need of a few mod cons.

'It's got a mini-bar with *cerveza* and that's what counts,' said Neil, and we sat on our veranda drinking beer and listening to the 'coo' of a blue-crested motmot, while the boys went off to check out the hacienda's small swimming pool. This was about as far removed from the all-singing, all-dancing resorts of Cancun as you could get; a place to regroup, take a breath and relax.

It also had one other unique selling point – a modest garden gate, the very same one that the first archaeologists had stepped through to explore, and it was opened for guests a whole hour before the official opening time of the site. At 9a.m. next morning we hurried through, keen to view one of the new Seven Wonders of the World – the 25-metre high Castillo de Kukulcan – before the tourists flocked in. Designed with mathematical brilliance to represent the Mayan Calendar, it's part pyramid, part clock, part fortress and wholly impressive. I'd arranged to meet a local guide there, who would give us a behind-the-scenes tour of the ruins.

'He's late,' I said, checking my watch, not yet familiar with the Mexican concept that thirty minutes past a scheduled meeting time is considered punctual. 'Twelve minutes late to be precise.'

We waited another ten before a cheery greeting of 'Hola' came our way. I turned to see a young guy with long hair pulled into a ponytail, wearing Ray-Ban sunglasses and a t-shirt that had 'Touched by Jesus' printed on it, and in much smaller lettering underneath, 'In a Mexican Prison'.

'Hi, I'm Jesus,' he said.

It's a name so common in Mexico that by the end of the trip we'd met several, but as he was our first 'Jesus' the boy's snorts of disbelief and amusement were hard to ignore.

'Sorry,' I said. 'Jesus isn't a very common name where we come from, unless you're the son of God.'

Jesus grinned and winked at me. 'Well, I'm not that well-behaved,' he said. Was Jesus flirting with me?

'Mum's gone red,' said Freddie.

Jesus marched us off briskly to the centre of an impressive oblong court known as the Juego de Pelota, where he ran around in circles. It was obvious to me that he ate a lot of chaya.

'You know basketball?' he asked the boys.

They nodded. In mentioning sport, Jesus now had them in the palm of his hand.

'Well, this is where the Mayans played basketball but with one very big difference,' he went on. 'Want to guess what that was?'

'Were the hoops smaller?' suggested Ben.

'The ball was smaller, the hoops were smaller, and there were lots of them hanging along the top of the walls, but that's not it,' he told them.

'Did you kick the ball?' asked Josh.

'Aha! Now you're getting closer.'

'So like basketball but using your feet?' Ben was always the one who sought clarification about things that sounded utterly ridiculous.

'Yes, they had to pass it to each other using everything but their hands and look how high the rings were.'

We looked up to the top of the court where the holes that held the stone rings could still be seen.

'That's impossible,' said Josh.

'Almost. The game ended on the first goal and often that took hours to score, but there was one other big difference,' Jesus told us. 'The captain of the losing team was sacrificed by having his head chopped off.' We watched Jesus run a slow finger across his neck.

'That's a bit harsh,' said Josh, master of the understatement.

'But if you were the winner you were treated like a God.'

'I wouldn't play,' said Freddie.

'Then you'd probably be sacrificed for disrespecting the elders,' Jesus told him. 'Want to see what happened to the losing captains?'

We followed him to where gruesome graphic images of decapitations were sculpted into the wall. 'You think that's bad, come with me,' Jesus said, leading us on to the Plataforma de los Cráneos, which was decorated with images of eagles ripping open human chests and feasting upon the hearts.

'This platform was used to display the heads of sacrificial victims,' Jesus told us cheerily.

At Cenote Sagrado (a scared limestone sinkhole) we stared into murky green water. 'Oh, you wouldn't believe the number of human remains they've dredged from here,' Jesus said. 'Human sacrifices didn't just involve having your head chopped off. Sometimes they tied rocks to your feet and made you jump into the water.'

There was a snack-bar just next door. 'Anyone like an ice cream?' Jesus asked. Even the boys had lost their appetite.

'I think I've had enough,' I whispered to Neil.

'Keep the faith, Kate. Now you've found Jesus, it would be awkward to lose him quite so soon.'

He was right. It wasn't Jesus's fault that beyond the Mayans' interest in all things astrological and mathematical, they were a blood thirsty lot. Decapitations, sacrifices, human remains, death, death and more death wasn't quite what I'd had in mind for lifting our spirits. Even so, we soldiered on. I zoned in and out of Jesus's running commentary, catching descriptions such as 'clawed his innards out' (something to do with a ferocious jaguar) and 'gorged his eyes out with his beak' (a story about an eagle with a grudge), before deciding enough was enough. We ended our tour with Jesus informing us that Mayans weren't averse to eating human flesh in cannibalistic feasts.

'Jesus, that was horrific,' I told Jesus, when our two-hour tour was at an end.

'Gracias,' he answered.

We decided to give Mayan history a rest for a few days and looked to Yucatan's natural beauty and wildlife to cheer us up. The 146,000-acre Celestún Biosphere Reserve, where thousands of flamingos flock annually to nest and breed, looked like the perfect place to start. I love these daft birds, at their most curious in the air when their necks and legs are equal distance from their wings, which makes them look like they are flying backwards. And talking of daft, when we sailed into the mangrove to get a closer look at the huge termite mounds that teetered on branches, we came across local people splashing in and out of a natural spring within snapping distance of a crocodile that was lounging on the bank.

'That's risky! Aren't they scared?' I asked the skipper of our boat.

'Oh no. They believe in what will be will be. Either the crocodile will eat them or not.'

It was such a different way of looking at things.

'But they'd be less likely to be eaten if they weren't in the water,' suggested Josh.

'If that is their death, then the crocodile will find them.'

'What even if you live in a high-rise flat in Mexico City?' asked Ben.

'They might work at a zoo,' suggested Freddie.

I could tell by Josh and Ben's eye rolling that they thought our skipper was talking nonsense. Having grown up with the Western belief that you were in charge of your own destiny, this acceptance of fate was a hard one not to challenge, particularly when, like Josh, you have recently joined the ranks of teenagers and, like Ben, you were on the cusp of becoming one. Scepticism aside, here we were in a boat on a lake discussing death again. I was beginning to appreciate the everyday intimacy Mexicans have with the subject, not so very different from our British preoccupation with weather. To them, it was just part of life. The more time I spent in Mexico, the more ridiculous the Western avoidance of talking about death was beginning to seem, let alone the ongoing and futile obsession many had in remaining forever young.

'I guess when your number's up, your number's up,' said Neil, and on that optimistic note we cruised back to the jetty. I would have liked to have found a positive way to continue the discussion, but by now Freddie was shrieking as Josh flicked the tiny pink shrimp that our skipper had scooped from the lake to show us what gave the flamingos their distinctive colour.

We stopped for a sand-between-our-toes lunch on the beach in Celestún, where families were barbecuing traditional Maya Tikin Xic fish (wrapped in banana leaves and chargrilled), their buckets of iced Sol beer buried in the sand. Hammocks were strung between poles out to sea and on small market stalls, conch shells and shark jaws were for sale. Freddie, always looking for a way to part with holiday pocket money, looked eagerly from the shark jaw to me. 'I'd never get it through customs,' I told him.

I knew it was only a matter of time before the subject of the *Dia de los Muertos*, the Day of the Dead festival, came up, and that happened two days later, as we found ourselves staring at a papier mâché figure in Casa de los Venados – a restored colonial house, decorated with more than three thousand pieces of Mexican folk art in the city of Valladolid. This gave me the perfect opportunity to broach the 'D' word again.

Among the collection were grinning papier mâché skeleton dogs and skulls wearing sombreros, but it was a cheeky-looking skeleton riding a bicycle and wearing a jaunty straw hat that had caught Freddie's eye.

'Do you like them?' I asked, pointing to grinning skeleton cat.

Freddie nodded.

'Mexicans like to poke fun at death,' I continued. 'It doesn't worry them at all. They even have parties in the graveyards every year during the Day of the Dead festival, so the people they love who've died don't miss out.'

'But they're dead,' said Ben.

'Not to the people who love them. The belief is the dead are given a one-day pass to return to see their families.'

'Not really?' asked Freddie, looking alarmed.

'Well, it's not like you see them,' I quickly explained.

'Sounds a bit creepy,' commented Josh.

'For them, it's a lovely way of remembering the people they love. They bring gifts and food to the graves, and if it's a young child who has died they bring toys. Mexicans believe that you're not really dead until the last memory of you fades.'

'When does this festival happen?' Freddie wanted to know.

'Oh, don't tell me that we're going to it,' groaned Ben, always a little suspicious of what madcap adventure I had planned for them next.

'Of course not!' Neil said. To be honest, though, if it had been November when the festival takes place, there was no way I would have wanted to miss out.

I saw Freddie relax. 'I do like these skeletons,' he said.

I mentally ticked a box. Death had been mentioned and in a positive way, associated with festivities. The idea took a bit of getting used to, but I could see that they were mulling it all over, even if they were a little freaked out by the idea of a party on a grave.

'Know who would have loved these figures? Rob.'

Rob had been an artist and had been full of fun. I could hear him having a chuckle over these works of art while appreciating the creativity that had gone into their execution. We carried on exploring, pointing out other things Rob would have liked – a clay pot in the shape of a toucan, a papier mâché pirate skeleton with a parrot on its shoulder, a rattan table woven to the shape of a monkey. I realized it was the first time we'd talked about Rob without being sad, which, with so many lovely memories of him to share, felt like a huge step forward.

Perhaps a coincidence, but while staying in Mérida, I noticed a change in the boys. While pedalling around the city on our rental bikes the next day, I listened as they called out a cheery 'Hola!' and tinkled their bells at everyone we passed, never failing to receive a friendly greeting back.

'I really like Mexico,' Freddie told me later, while strumming on his guitar. I bent over and gave him a kiss. Easy-going, eager to please, he was the third child we could have easily missed out on if we'd settled for a less chaotic life with two kids, and we all loved him to bits. Minutes before, I'd seen the delight on Josh and Ben's faces as Freddie clowned around while singing and playing his guitar. This moment of sun-filled pleasure on the terrace of our little hotel had made me teary. This was the happiest I'd seen the boys in months.

On our last day in Mérida, we took a trip out to a nearby *cenote* (sinkhole), just one in Yucatan's vast underground network of caves, subterranean rivers and sinkholes. At 20 metres below ground level, and partially open to the sky, access to it was by a slippery moss-carpeted wooden staircase on one side. Vines and small waterfalls cascaded down the other and in the natural pool at the bottom we floated on our backs alongside the catfish, shrieking at every stroke of passing fin. Life suddenly felt good again. A weight had been lifted – I could hear it in the boy's laughter and feel it myself, as my shoulders dropped five inches and I let the tension I'd been carrying with me for months ebb away.

On our drive back into the city, the traffic suddenly slowed.

'What's happening?' asked Ben.

I blinked into the sunlight from the back of the minibus. The midday sun bounced off the shiny exterior of a bashed up red VW Beetle in front of us. 'I can't really see,' I told him.

'Funeral,' our driver told us. 'Want to get out and take a look?'

Never in a million years would you be asked that question in the UK, but already the suggestion felt normal. He pulled the car over and we jumped out.

A little shyly, we joined the crowd. More carnival procession than funeral march, a mariachi band followed the priest at the front. Family members and friends escorted the coffin, chatting and smiling. I knew that the outpouring of grief would come later at the grave.

'Perhaps he was a musician because look at the flowers,' I told Freddie, pointing to where, on top of the coffin, a floral arrangement was in the shape of a perfect guitar.

'And is that a picture of Elvis on the side of the coffin?' Josh wanted to know.

I grabbed Freddie's hand and we crossed the road to get a better look. On closer inspection, we decided that it wasn't Elvis, but actually a picture of the deceased, who could easily have passed for Elvis's Mexican brother with his jet-black coiffed hair and sideburns.

Was it strange to be following the coffin of a man we didn't know, discussing his hobbies and hair style? Perhaps a little. Was it sad or disturbing? Not in the slightest.

'You could have flowers in the shape of a wine bottle,' joked Josh to me.

I heard Freddie stifle a giggle.

I think it was true to say that Mexico had done the trick. While we weren't perhaps laughing in the face of death, the subject was no longer taboo, and even extended to my sons making jokes about my own demise. That might take a bit of getting used to.

'I'd love that,' I replied, meaning every word.

11

Morocco

A Lesson in Not Judging a
Book by its Cover

Josh age fourteen, Ben thirteen and Freddie nine

23 July 2015,
Fez, Morocco

Interesting moment at lunch today when Neil ordered a camel burger in Café Clock, which had been recommended to us by a Swedish tourist we'd met in the souk. 'It's a little, how do you English put it, bonkers?' she'd said, and she was right. Slap bang in the middle of the medina, we sat at a table sandwiched between a grandfather clock and a shop mannequin dressed in a Berber robe; above us was a chandelier made of slender brass hunting horns.

'What are you having?' I asked the boys, already guessing that it would be a request for scrambled eggs, the blandest item on the menu.

'I'd like the camel burger, please,' said Ben to the waitress.

I gave Neil a swift kick under the table to stop him asking if Ben thought that was really such a good idea. I couldn't wait to see how this would pan out.

'Yeah, me too, please,' said Josh.

'And me,' chimed Freddie.

I knew there was a good chance that they thought 'camel burger' was just a cute gimmicky name referencing the country they were in, rather than a burger made of actual camel meat.

I ordered the falafel. No camel for me. Dirty, smelly, spitting creatures that they were, with their rubbery lips and fly-encrusted eyes.

Behind the boys, twelve iconic red cylindrical Fez hats with tassels were arranged in a diamond shape on the wall.

'How do you get three fussy boys to eat camel in Fez? Just like that,' I whispered to Neil in my best Tommy Cooper voice.[19]

When the food arrived, I braced myself.

'What's that stuff?' Ben asked, pushing a splodge of something red away from his burger. That's the least of your worries, I thought.

'It comes with something called taza ketchup,' I told him, reading from the menu.

'Hmph,' said Ben.

As in countless other places, I watched Ben make the (futile) request to the waitress for Heinz tomato sauce (their optimism on that score, from Borneo to Morocco, would never die).

---

[19] Tommy Cooper (1921–1984) was a British comedian and magician, who wore a red Fez hat while performing and whose catch phrase was, 'Just like that.'

Ben was the first to take a bite of camel. 'I wasn't sure when I ordered it but camel tastes good,' he said

Un-bloody-believable. I can't get Ben to eat a fish finger at home, but in Morocco he is eating camel.

'It's not really camel, is it?' asked Freddie, while sniffing suspiciously at the burger.

Josh was quiet. It was obvious that, like Freddie, he too had been duped. But he'd never hear the last of it if he let Ben take all the camel-eating glory. Oh happy, happy days as I watched him take a bite and chew slowly. When I asked him if he liked it, he screwed up his face. 'It's alright,' he said. 'But I think they should make it clearer on the menu that it's made with camel.' I resisted pointing out that the clue was in the name.

Neil egged me on to take a bite. According to him, it tasted 'just like beef only a little more gamey' (having just googled 'camel meat', I've learned that the gamier the taste, the older the camel. Vomit.)

As I spend eighty per cent of my time trying to cajole the boys into trying new foods, I hate it when I'm put on the spot like this. 'I didn't order it, so why should I?' I snapped.

'Oooooo no need to get the hump about it,' Neil said.

~~~

Of all the places we'd taken the boys to, I was the least sure about Fez. Telling them that the UNESCO World Heritage Site Fez el-Bali, the ninth-century medina, was the largest urban car-free zone on the planet (I always resorted to the word 'planet', with its exciting Star Wars connotations, than boring old 'world' when I was trying to talk something up), had drawn a tiny glimmer of interest, but they still weren't exactly thrilled at the prospect of four days there. I, however, was itching to visit, knowing that now was the time, while the walls still crumbled, and the mules and donkeys weren't afraid to fart in the face of tourists. I liked the idea of stepping back in time to a place unchanged for centuries and there weren't many places as authentic as Fez left to explore. It hadn't helped that the boys had overheard me telling Neil I'd been invited to review a new five-star hotel in Walt Disney World Florida, which I'd passed up because, to be quite frank, I'd rather pull my teeth out with pliers than endure that utter hell.

'What? You've said no?' asked Ben.

'Oh, Mum. Please, can we go?' groaned Freddie.

'How could you?' asked Josh.

'I don't want to go,' I said. 'I want to go to Fez.'

'Fez? Who the hell has even heard of Fez?' asked Josh.

'Exactly! It'll be really interesting,' I told them. 'And now's the time to visit.'

'Yeah, but does it have *It's a Small World*?' Neil teased me, referring to the ride that drove me to the brink of insanity on a trip to Euro Disney.

Ignoring him I continued, 'It's like Marrakech was a hundred years ago.'

'Marrakech a hundred years ago?' repeated Josh, looking utterly gob-smacked, as if I'd just told him that I wasn't his birth mother (as he wished right now). 'What! Older and worse than Marrakech?'

We'd taken the boys to Marrakech in 2007, and Josh had been draped with a terrifyingly large snake the minute we had stepped into Jemaa el-Fnaa, the main square. As I recollected, though, that had bothered me more than him. 'What was wrong with Marrakech?' I demanded.

'The food! Remember the sheep heads!' Josh said. Ah, yes. I'd forgotten about that rather alarming sight but clearly the memory of them boiling away in pans at the food market was still haunting him.

'We ate sheep heads?' asked Freddie. He would have been only two at the time so relied on his older brothers to fill in the gaps.

'For God's sake, I wouldn't give you a sheep's head to eat, would I?'

Freddie didn't look sure.

We needed to get off this subject and fast. I was good at finding compromises and already had one planned for this trip, which I pulled from my sleeve right away.

'After Fez, we're going by train to a beach resort near Casablanca where you're all going to take part in a PSG football academy.' Straight in the back of the net. In the words of commentator Kenneth Wolstenholme during the 1966 World Cup Final, when England beat West Germany 4–2, 'They think it's all over! It is now!' I heard not one more complaint.

Looking back, July might not have been the best month to visit Fez with an unwavering temperature of 36 °C nuancing every smell in the packed alleyways of the old medina where we were staying.

'Bit smelly here,' said Freddie as we hauled our luggage through a narrow, cobbled street, too small for the taxi we'd taken from the airport. We were in search of the *riad* (traditional house built around a courtyard) turned boutique hotel that I'd booked for our four-night stay, which we eventually found hidden behind a heavy wooden antique door. In the centre of a large tiled courtyard a fountain burbled, and the *riad*'s stained-glass windows, illuminated by the midday sun, sent a kaleidoscope of colour across the carved stone walls of this beautiful Moorish house. The air was perfumed with jasmine, orange blossom and mint, and, for me, it was like stepping into paradise. The boys, however, looked less impressed.

'Is that the pool?' asked Freddie, pointing to where the fountain flowed into a large ornamental pond that ran the entire length of the courtyard.

'Don't even think about it,' I said.

Amid potted palms, a tourist sat sipping mint tea, looking effortlessly cool in white linen (damn her), which was a look I'd tried to emulate several times but never quite mastered with three grubby-fingered sons. Some years earlier, while wearing a white sundress, I'd sat on an abandoned chocolate gelato and then spent three hours walking around Rome, listening to Josh and Ben chant, 'Mummy's pooed herself.'

After checking in, we stared at a map of the medina and listened as the receptionist gave us directions to Bab Bou Jeloud, the medina's famous blue-tiled gateway, which she suggested would be a good place to start our exploration.

Out on the street, Neil and I looked left, then right. 'Get any of that?' I asked him.

He shook his head. So, I folded the map up and put it in my bag where I knew it would stay.

'Let's just wander, shall we?' I suggested. There were nine thousand tiny streets in the old city alone. Expert navigator that he was, even Neil looked overwhelmed. Predictably, we were lost after just ten minutes.

'I feel that we should take a right up there,' I said.

'Which would take us back to the hotel,' Neil told me. 'That's the way we've just come.'

We spent much of our first day walking in circles with Neil stepping in mule and donkey excrement at regular intervals of five minutes in what Ben was now referring to as 'Dad's shit-detector shoes'.

'I feel sick,' Josh said, slumping down to sit on a stone step.

'Really?' I asked. Was it the discovery, before we'd set out, of a lovely swimming pool tucked away in the *riad*'s palm-shaded garden that had prompted this sudden malaise?

'Yeah, I think I might throw up,' he told us, worryingly doubling over.

We were trying to locate Souk Sabbaghine, the Dyers' Souk, where I'd read that streams of rainbow coloured water ran down the streets to the river. 'Oh, you'll be okay,' I said, just as Josh spilled the remains of his camel burger on to the ground.

A rotund local lady, dressed in black and carrying a bucket of water, appeared from a hole in the wall and swilled the vomit away. She tapped me on the arm in a reassuring, one mother to another kind of way, which seemed to highlight all my failings (number one being that we were not queuing for Space Mountain in Walt Disney World but instead watching Josh retch within a foot of a pile of fresh donkey crap).

'Feel better now?' I asked hopefully.

'I just want to go back to the hotel,' Josh said miserably. 'You carry on, though. I can find my way back.'

'What, on your own?'

'I'm fourteen, Mum,' Josh said. 'We came this way this morning. I'm pretty sure I know my way back from here.'

'I'll go with him,' offered Ben.

This place was like the Crystal Maze.[20] There was no way I was letting them wander off by themselves.

'Actually, I think I've had enough for one day anyway,' I lied. 'Let's all head back.'

There were times that you just had to accept while travelling with kids that the day was not going to go as you'd planned, and this was one of them. While Josh took a nap and the others went to swim at the pool, I sat on our roof top terrace with a swift's-eye view of the medina and beyond to the Atlas Mountains. The aroma of roasted almonds rose up from a street vendor's cart and the rousing call to prayer could be heard from a nearby mosque, and I was starting to think that the trip had been a bad idea. I'd travelled to places such as India, Nepal and China, but had never felt quite so out of my depth. It was clear we needed a guide.

We decided to have dinner in the *riad* that evening, mainly because stepping foot into the medina's labyrinth in the dark would have been completely nuts, speaking of which . . .

---

[20] *The Crystal Maze* is a British game show, in which contestants are locked into a labyrinth, confusing in terms of both time and space, and must complete a series of challenges to escape.

'Has the beef tajine got nuts in it?' I asked the waiter. Surely we were on to a winner here.

'Oh, yes. That has no nuts apart from the almond.'

'That's a nut,' I told him wearily.

'What about this?' Neil asked, gesturing to a dish described in English as lemon chicken with olives.

'Excellent choice.'

'And nut free?'

'Yes, apart from the almonds.'

We'd had conversations like this before in other parts of the world. It is a mind-blowing concept in so many cultures that anyone could be allergic to a nut, and even when I waved an EpiPen around and pantomimed stabbing it into Josh's leg, they still didn't seem to quite get it. Nuts were a staple in Moroccan cooking and almonds, in particular, were proving impossible to avoid. Roasted, grated, chopped and sprinkled – they showed up in the most unlikely recipes, so we needed to be on our toes. I showed him the translation again. It said, as usual, that Josh had a life-threatening allergy to nut and nut oil, and the waiter nodded like he understood. Then flipping to another page in the menu, pointed to a flaky pastry filled with shredded chicken and wait for it . . . almonds.

'I'm not really that hungry to be honest,' said Josh, who, although feeling much better, still had a rather nasty case of what he was calling camel burp. 'Can I just have some vanilla ice cream?'

The ice cream arrived sprinkled with chopped pistachio. 'I'll eat bread,' said Josh.

Next day, I decided that we should start afresh. 'Let's pretend this is our first day here,' I told the boys, as we waited for Rafik, the guide I'd booked for the morning.

'Assalamu alaikum, (peace upon you),' Rafik called as he entered the riad's courtyard. His shaven head was topped with a Fez hat and I could tell instantly that he would be fun.

'Yesterday, we found everything a bit daunting,' I admitted, when he asked me what we'd seen already.

'Ah, it's difficult if you're a tourist. I see things I've never noticed before each day and I grew up here. But the most important thing to remember is, how do you say, never judge a book by its cover?' he asked me, disappearing into the narrowest of alleyways. Like a Moroccan Willy Wonka,[21] Rafik danced down small side streets, disappearing around corners only to reappear with a grin whenever we thought we'd lost him. As we walked, he pushed open doors to reveal all sorts of surprises.

'Behind many humble doors in these simple streets there are beautiful houses with gardens full

---

[21] Willy Wonka is a mischievous, magical, fictional character created by the author Roald Dahl in his 1964 book, *Charlie and the Chocolate Factory.*

of flowers,' he told me. 'In Fez, appearances can often be deceptive.'

We peeked into a lemon tree filled courtyard where a mother washed her baby in a tin tub and into a palm-shaded terrace where two old ladies sat splitting peas. Behind one door, we found a knife grinder turning his stone wheel; behind another, a weaver shuttled a bobbin across a loom. It was like opening a human advent calendar.

'Can you smell something?' Rafik asked the boys.

Ridiculous question to ask Ben, who, with the most sensitive nose of us all, had his pegged tightly shut already.

'It's not nice whatever it is,' said Josh covering his too.

'The smell is for free. But the leather is not,' Rafik joked, giving us a clue to where we were headed. From a market stall, he stopped to buy fresh mint and gave us each a sprig. 'Best to hold to the nose,' he helpfully said.

'It's like being in the bat poo cave in Borneo all over again,' Ben muttered. He was right. A pungent smell of ammonia filled the air.

The tanneries in Chouara are Fez's most Medieval sight, where a vast honeycomb of vats is used for treating and dying the animal skins, and cauldrons of pigeon dung are used to soften the animal skins.

'It's like we've stepped out of the Tardis,'[22] said Neil as we gazed down on it all from the second-floor balcony of a leather shop. Once upon a time that *Doctor Who* analogy would have worked a treat on our sons. We could have played a game in which we were time-travellers, transported back to a world where we were knee deep in bird shit – far worse than Daleks, Cybermen or anything else that the Doctor had encountered. Now we only had Freddie to humour us.

'Isn't there an easier way to treat the leather?' I asked. 'Breathing in this smell all day can't be good for the people who work here.'

'Tradition is important in Fez,' explained Rafik. 'For a while there was talk of modernizing and moving the tanneries, but in the end it stays the same.'

'For the tourists to see?' I wondered.

'Tourists are welcome but not essential to our way of life. You'll see. I want you to meet a friend of mine.'

Back on the street we followed Rafik through a warren of back alleys to a small row of workshops, which each measured around 3 metres square. 'Here is Nizar, my neighbour and good friend,' said Rafik,

---

[22] The Tardis is a time machine and spacecraft used by the Doctor in the British science-fiction television series *Doctor Who*, which first aired in 1963.

reaching down to clasp the hand of an old man who sat cross-legged on a mat, carving a comb. Sparse of hair and teeth, with skin as wrinkled as a raisin, Nizar looked like he'd been around the block a few times.

'Salam!' we all chorused.

'Salam! You are welcome!' Nizar told us.

'Nizar makes combs from cattle horns and he's been here since 1952. His father worked here with him until the day he died.'

To be honest if Rafik had told me Nizar had been there since the ninth century I may well have believed it. 'If he doesn't mind me asking, how old is he?'

After an exchange of words with Nizar and some head scratching, Rafik said, 'He thinks he is eighty-four, but maybe older.'

'And how many combs does he sell a day?' What I was really wondering was how on earth Nizar could make a living.

'He makes them for the people who live in the medina, for whenever they need one,' Rafik told me.

With a three-toothed grin, Nizar held out a small yellow bone comb that he had fashioned into the shape of fish.

'Can I have it?' Freddie asked.

'Of course,' I said, reaching into my bag for my purse.

Nizar shook his head. 'It's a gift,' Rafik said.

It was one of those moments while travelling that catches you by surprise. 'Oh, that's so kind,' I said.

Here was an old man eking out a living and he'd given up the chance to make a few extra dirhams from a tourist. He seemed to embody what Fez was about – a city of tradition and honour that existed and upheld its way of life for the people who lived there, not for tourists to gawp at, and Nizar's gesture had moved me.

'Thank you,' said Freddie, combing his scruffy mop with it, which made Nizar's smile beam even brighter.

'Aww, he was nice,' said Josh, as we followed Rafik through the souk. 'Has he got grandchildren?'

'He's got a son your age,' Rafik said.

'A son?' we all screeched.

'Yes, his second wife is forty years younger.'

It wasn't just the humble doors of the houses in the medina that concealed surprises.

Rafik had been right about not judging a book by its cover. Nizar's cover looked like it should have been in the bag for the charity shop, not filed under erotica.

When we parted ways with Rafik he gave us this advice: 'If you get lost, walk uphill. The medina is shaped like a bowl, so at some point you will find a way out.' Then he hurried away down an alleyway glancing at his watch, now looking more like the White Rabbit from *Alice in Wonderland.*

'Pool?' suggested Neil.

'Okay if Ben and I look around the souk?' Josh tentatively asked.

It was obvious a plan had been hatched, one that involved cutting their mother's apron strings. I tried to play it cool. 'Oh, what's in the souk?'

'T-shirts,' said Josh.

'Fake footy shirts,' added Ben.

'We could all stay,' I suggested.

'Nah, you're alright,' said Josh.

'Well, we'll wait for you in a café while you pop in,' I said.

'No need,' said Ben.

'It's only a ten-minute walk back to the hotel,' Neil reminded me.

A ten-minute walk where our sons could be mugged, murdered, drugged, kidnapped for a ransom, trampled by mules, hit by falling debris from a Medieval building, washed away in a flash flood from the Dyers Souk, force fed almonds . . .

'And what's the worst that could happen?' Neil asked. Sometimes, he really did have no bloody idea. Reluctantly, I agreed.

Thank God, for the *riad's* lovely pool because by 3p.m. each day we were beaten by the heat. No breeze seemed to make it past the thick stone wall

that wrapped the old city in a hug and the air in the
medina's small alleyways was stifling.

'Stop worrying,' Neil told me, as he caught me
checking my watch for the umpteenth time.

It had only been forty minutes, but it felt like more.
'I'm giving them another ten minutes, and then one
of us will have to go and find them,' I told him. By
one of us, I meant Neil.

'No need,' Neil said, nodding over to where Josh
and Ben were sauntering towards us through the
gardens. Playing it cool, I quickly rearranged my face
from extremely anxious to totally unbothered.

'Buy anything?' I asked them.

'Nah. Too hot for shopping,' said Josh.

Empty-handed as they were, I realized that the
exercise hadn't really been about going to the souk,
but more about the freedom of being allowed to go
there if they so wished. They were growing up that
was clear, however, in so many other ways, they were
still young boys. As if to confirm this thought, I heard:
'Body bomb,' from Ben as he jumped into the pool.

'Cannonball,' yelled Freddie as he too tucked into
a ball to enter the cool water.

On the other side of the pool, the lady in white
linen (now called Araminta in my head) who I'd
spotted on the first day, was lying on a sunbed
flicking lazily through a magazine. She wasn't
wearing white today. Instead she had on a red

Chanel bikini (damn her), enormous dark shades and a wide brimmed floppy hat. She was a reminder of the number one thing that I missed the most since becoming a mother – having time to myself (Araminta looked like she put her time to good use in coordinating outfits for the pool, whereas I would most probably have squandered it on my favourite of pastimes, daydreaming). Nearby an older lady sat reading a book in the shade of a palm.

'Mum, Mum, Ben held me under, and I couldn't breathe,' shouted Freddie.

'You don't seem to be having a problem now,' I replied, still with my gaze on Araminta, who was spritzing her face.

'Ben, Freddie, look at this,' shouted Josh, as he did a back flip into the pool.

Araminta leapt to her perfectly manicured feet and began wiping non-existent water from her legs. 'Would you mind keeping your children under control?' she trilled at me.

'Boys please be careful,' I shouted. 'Sorry about that.'

I watched as the boys swam to the opposite end of the pool, away from Araminta. 'Okay?' I asked her.

'No, not really.'

'Oh?' I replied, rising from my sunbed and sucking in my stomach. 'Why, what's the problem?'

'The problem is that this hotel isn't suitable for children. You shouldn't be staying here.'

'Excuse me?' I felt my hands go to my hips, an indication to Araminta that this was now war.

Neil tended to leave the argumentative stuff to me, but on this day Araminta had pressed a button. 'If you've got a problem with us, I suggest that you go and talk to the hotel manager,' he said calmly but firmly. 'But in the meantime, as paying guests, my sons have every right to play in this pool.'

'Well, really. How dare you talk to me like that. That's exactly what I'll do,' she shrieked, slipping into a stunning, what I suspected might be a Missoni (damn her) kaftan, before storming off. 'Mummy, are you coming?' she barked at the older lady, who looked wearily up from her book.

'No, dear. I think I'll stay here,' she said.

When Araminta had gone, her mother wandered over to us. 'So sorry about my daughter. She was jilted at the aisle two days ago. This is her honeymoon.'

Ahmed's words rang in my ears again, 'Never judge a book by its cover.' Araminta, who turned out to be called Anabel (close enough), was another fine example of how appearances could be deceptive.

'We know a nice man who makes combs in the market,' Neil whispered to me. 'Perhaps he's got a brother.'

12

Tanzania
A Lesson in the Birds, the Bees, the Monkeys & the Zebras

Josh age fifteen, Ben fourteen and Freddie ten

6 August 2016,
Serengeti National Park, Tanzania

When I suggested to Josh and Ben that they might like to join Freddie for something called A Walk on the Wild Side, organized through the lodge's kids' club, their mouths fell wide open.

'Are you kidding? I'm far too old to go to kids' club,' Ben told me.

'And if he is, then I definitely am,' said Josh.

'Oh, I just thought you might like to spend a couple of hours with Lemuani and the other Masai guys. It's them who are taking Freddie out,' I said, as casually as I could.

I saw Ben glance at Josh. The Masai are employed by the lodge as trackers, guides and security guards, and in the coolness stakes they are up there, particularly one young spear-carrying warrior called Lemuani, who likes to chat about English footy with Josh and Ben.

'What's Freddie doing with them?' Ben wanted to know.

'Bush skills. How to spot lion tracks, that kind of thing.'

I tried not to laugh. Fourteen and fifteen are tricky ages, no longer a child, but not yet a man. It was clear by their faces they were regretting their hasty response.

'Okay, we'll go. Just to keep Freddie company,' said Josh.

Two hours later they were back and although Josh and Ben played it cool, I could see that all three had enjoyed themselves.

According to Ben it had been like an edgy scout trip, and they'd all had a go at whittling a toothbrush from a twig with Lemuani's knife. And to think that I used to worry about them eating play dough at nursery.

'What else does he do with his knife?' I asked.

'All sorts of cool stuff,' said Freddie, who was never into the finer details.

As I listened to them telling us about the camera traps they'd set on the perimeter fence of the lodge to record any animals that might visit by night, it occurred to me that the Masai guards weren't there to add authenticity to our experience by teaching the boys how to make twig toothbrushes, but rather to protect us should anything go wrong. That spear, that knife – they were carried for purpose.

I made the mistake of suggesting that we should google 'when things go wrong on safari', and we spent the next half hour reading horrific stories about (bloody stupid) tourists who had stepped out of their safari jeeps to take close-up *pictures of lions and had been mauled to death and eaten. Not to mention being gored by a rhino, ambushed by a hippo and possibly the most bizarre, having your safari vehicle humped by a randy young bull elephant. Even those gentle giants, the giraffes, had been known to kick tourists into touch.*

'Who would be eaten first out of us?' wondered Ben.

'Probably Dad. There's more of him,' I said.

'But what if the lion had already eaten and just fancied a snack,' asked Josh, and we all looked at Freddie.

~

Having trialled a mini safari a few years earlier in Sri Lanka, our sons were now at an age that a longer safari would make for a great family holiday: old enough to behave appropriately, knowing when to be quiet so as not to scare the animals away, and interested enough to listen attentively to our guide. Most importantly, I felt that they would now fully appreciate how privileged they were in being given the chance to spend time with wild (often endangered) animals in one of the world's most incredible environments – the Serengeti National Park. I'd booked a private jeep and guide, so that we could suit ourselves in deciding which animals to track and how long to spend at each location. Some years ago, I'd spent five days on safari with an elderly couple from Idaho (Al who favoured bright yellow golfing slacks and Goldie who wore a pink neon sun visor) and I wasn't keen to risk a repeat of the experience. There's nowhere to hide in a safari jeep, and Goldie's constant cry of, 'Oh my gaaaaawd. That is just so adorable!' had driven me up the wall.

To reach the Serengeti had felt like an epic journey (rendering the usual question of 'Are we nearly there yet?' laughable) and the last leg, from Arusha in Tanzania to the National Park, in an eight-seater propeller plane, took us on a hair-raising roller coaster of a ride over volcanic craters and peaks. 'Sorry for the bumps,' the pilot called back to us. Although, the turbulence wasn't bothering Freddie, he was so exhausted from our long journey that he had been bounced to sleep. Over Kilimanjaro, with its summit shrouded in cloud, we flew, until we dropped in over the Serengeti, where I woke Freddie up in time to see the giraffe and herds of zebra that galloped and scattered as the plane came into land. At the airport (a dusty strip and wooden shack) we were met at arrivals (a wooden bench) by Priscus, our guide, who high-fived the boys and said, 'Shall we go and find some lion?'

'We won't really see lion just wandering around will we?' Freddie whispered to me, and I remembered that how, on your first safari, it takes a while to grasp that potentially dangerous animals really are just roaming free. My sons had only ever seen a lion behind the bars and electric fences of a zoo enclosure and today it was only a matter of five minutes before we spotted three lionesses lazing in the sun. 'Bloody hell, I thought he was joking,' said Josh.

En-route to the Lodge, we caught sight of a cheetah sauntering through a herd of Thompson's gazelles, warthogs trotting through the bush and zebra cantering across the red dust road. 'Congratulations boys!' said Priscus. 'You've just seen your first real zebra crossing.'

'Swim?' I asked the boys, after checking in at the lodge. I knew I had to keep them going for a few more hours before we all crashed out. Jet lag could ruin a holiday and we'd have to be up at the crack of dawn for our first full safari drive the following day. Also, I hadn't let on but I knew there was a good chance we may see elephants, which often came to bathe and drink at a watering hole that lay just below the lodge's swimming pool.

'What's that noise?' asked Ben, as we were lounging poolside.

I held my breath. Could it be? Every sound in the bush is amplified, from a mosquito's buzz to the heavy footfall of elephants, which was what I hoped we were listening to now. An elephant's call can carry for up to fifty kilometres, and when that distinctive trumpet filled the air the boys scrambled off their sunbeds.

'Was that an elephant?' Josh asked.

'Let's wait and see,' I told them, but we didn't have to wait long before the first came into view. There is nothing that can prepare you for the sight of fifty-three (Freddie counted them) elephants swaggering through the trees towards you. We all looked at each other and laughed at the luck of what we were witnessing.

'Take my picture with them,' shouted Ben diving into the pool. Behind Ben, in the frame, I could see juvenile elephants wading into the watering hole to drink and play, while younger ones kept close to their mums as they drank, perhaps fearful of falling in. The image I captured of Ben was so breathtaking, it looked photoshopped.

'Something in your eye?' asked Neil, as he caught me wiping away a tear. A question that was soon to become the joke of the holiday.

Excited as the boys were to be on safari, after our epic journey to get here and battling with predicable jet lag, rousing them before dawn for our first proper game drive wasn't easy.

'It's not my fault that this is when the animals are most active, so get up,' I shouted on my fourth attempt.

'Mum, go away! Let me sleep!' Josh begged.

Ben remained corpse-like, while Freddie rolled into a hedgehog-like ball and refused to even open his eyes. I resorted to tearing off the covers.

'Don't you know that teenagers need twelve hours of sleep,' moaned Ben, as they trooped zombie-style to our safari jeep, where Priscus was waiting.

'*Jambo* (hello),' he shouted.

'Sorry, we're a few minutes late,' I said. 'It was difficult getting the boys out of bed.'

'*Hakuna matata*,' replied Priscus.

'That means no worries,' Freddie informed us, looking pleased with himself.

'Huh?' said Neil, and I realized that he hadn't spent months of his life, as I had, watching Disney's *The Lion King* on a loop. He looked a bit surprised that Freddie was fluent in Swahili.

'It's what Timon, Pumbaa and Simba sing,' Freddie said.

'A meerkat, warthog and lion cub in *The Lion King*,' I explained.

'Oh,' Neil said. 'I remember.'

I was pretty sure he didn't, so sang a bit of it for him.

'Oh my God, it's too early to listen to this,' muttered Ben, disappearing under one of the blankets Priscus had just handed us to keep out the chilly morning air.

We bumped along the dirt track away from the lodge and into the National Park, and Priscus, used to being up at this hour, called back jokes to the boys as we drove towards the river to watch the sunrise.

'Want to know why the warthog's tails are so straight in the air? They're looking for Wi-Fi.'

'What's worse than a lion chasing your safari vehicle? Two lions chasing your safari vehicle.'

'What happened when the lion ate the safari guide's joke book? He felt funny,' Ben said from beneath his blanket. When it came to cheesy jokes, he had a stock of those himself.

In the first light of dawn, Priscus called our attention to the call of a ring-necked dove that sounded exactly like it was telling us to 'drink lager', and when an impala with a black 'M' on its rear crossed the road, he quipped that it was the nearest thing Tanzania had to McDonalds. Then suddenly he cut the engine. 'Up ahead,' he whispered. On the side of the track a female adult cheetah and three young were sitting like they were waiting for a bus. After working as a national park guide for ten years, Priscus surely must have seen it all and I realized how rare this sight must be as he reached for his camera.

'Unusual to see them so close to the road,' he told us. 'This is going to be a great day. The first animal you see is an omen and seeing cheetah is a good one.'

The cheetah sloped off, followed by her cubs that rolled in the grass to play-flight whenever their mother stopped to sniff the air. They reminded me of three other boys I knew, who also liked to mess around when their mum wasn't looking.

As the sun rose, the sky turned from an inky blue etched with violet to a vibrant tangerine. Priscus pulled up to the riverbank and parked at the edge; a little too close for my liking. I knew that we were in capable hands, but there were fifty or more hippos just below us that we'd be joining if the handbrake wasn't securely on.

'Aren't we a little too near to the edge?' I asked, making the mistake of adding a (nervous) laugh to keep my concern sounding light because all that Priscus, Neil and the boys did was chuckle right back at me.

The hippos snorts and grunts filled the air as they jostled for position, and the young ones, keen to wrestle, opened their mouths like flip-top lids to lock jaws with opponents.

'Just play fighting,' Priscus told us. 'In a real fight, it gets brutal.'

Nile crocodiles looked on from the bank opposite us, biding their time. If one of the young hippos (or a safari vehicle full of tourists) was foolish enough to fall towards them, they'd be ready.

'Shall we move on?' I suggested.

The boys, warmed by the sun and Priscus' engaging banter, were now on their feet with binoculars at the ready, and from the moment that

Ben spied a lilac-breasted roller's purple feathers glinting in the sun, it was game on to spot the next new animal or bird. I watched as Josh and Ben elbowed each other for space as we continued along the dusty park tracks and, when a pride of lions breakfasting on an unlucky zebra came into view, I braced myself for the debate over who had seen them first. Josh quickly claimed it had been him.

'I saw them at the same time,' argued Ben.

'Yeah, but you didn't say,' said Josh. 'So, it doesn't count.'

'Who says it doesn't. You don't make the rules!'

'Wait a minute,' said Josh. 'Who are they?'

While they'd been quarrelling another safari jeep, containing a merry band of Australians, had sidled up to us to have a look at what we'd found.

'G'day,' one shouted over. 'Ripper find, boys! Good on yer.'

Ben glared at them. 'Piss off, these are our lions,' I heard him mutter.

'Yeah, go and find your own lions,' Josh agreed, and just like that they were back on the same team.

Remaining in the jeep, we stopped for a picnic under the shade of a baobab tree and as I was busy doling out the pastries and juices that the lodge had packed for us, a vervet monkey boldly dropped through the open roof and landed next to Freddie.

'Don't move,' Priscus said.

No danger of that. Freddie was rigid with fear. The monkey rifled through the picnic box, rather surprisingly tossing aside a banana.

'What do you call a vervet monkey with a banana in each ear?' asked Priscus. 'Anything you like. He can't hear you.'

The boy's shoulders shook in silent laughter. The monkey, having got what he wanted (an apple), climbed out, but not before giving us a spectacular view of his bright blue testicles. However, they were no means the most impressive genitalia that we were to encounter that day.

'Bloody hell,' said Josh. 'Have you seen the size of that zebra's . . .'

'Yes, alright, Josh. No need to shout about it,' I said. Although it was hard to miss. Never mind hung like a horse. The expression surely should be hung like a zebra.

'It's enormous. Like another leg,' said Neil.

'It's practically dragging on the floor,' added Ben.

'Is that normal for a zebra?' I tentatively asked, wondering if we'd stumbled across a freakishly well-endowed porn star kind of zebra.

'Oh, yes. The penis can be up to a foot and a half in length. The longest of the horse family.'

Ridiculous to think for a second that Priscus would not be able to provide us with facts on Zebra genitalia, but I was little prepared for what came next. 'But it's the female Grévy's zebra that is the most impressive with what we call a clever vagina,' he began.

Suddenly everyone was quiet. Priscus had just dropped the 'V' word. There was an unwritten, unspoken rule in our house – my sons could make as many willy jokes as they liked but they were never EVER to make references to the female equivalent. You could cut the atmosphere with a knife.

'How do you mean?' I asked. Obviously it was up to me to inquire more on this subject. Not even Neil would have dared.

'She can dump the sperm of a male who has disappointed her,' Priscus explained to us, grinning a wide smile.

I couldn't resist a little chortle. 'I might regret asking this but how exactly?'

'Very strong muscles. She can pump it out very fast. Even before the male has finished.'

I glanced at Ben, who'd tucked his neck into his shoulders, like a turtle trying to disappear into its protective shell.

'Okay, I think we'll leave it at that,' I told Priscus.

I'd been prepared for gore. We'd already encountered a gazelle ripped to bits, it's skeleton eerily hanging from a tree after a leopard had hauled it up there to feast upon; and the carcass of a rotting hippo slaughtered by a crocodile, then picked over by vultures; and the entrails of a dik-dik (a type of small gazelle, and a favourite with the boys for obvious reasons) smeared across a lion's chops. Josh, Ben and Freddie appeared to take it all in their strides, understanding the food chain and survival of the fittest. What I had, stupidly, forgotten about was all the fornication that goes on, in its most uninhibited, full-on grunting, don't care a hoot if there's a safari vehicle full of people taking pictures, kind of way. If I thought that a Grévy's zebra's clever vagina was the worst it could get, I could think again.

'The juvenile males like to dry hump each other. It's practice for later,' Priscus informed us as we watched one excited male vervet monkey grind against the other in an act of non-penetrative sex.

I heard Josh and Ben guffaw, while Freddie kept his eyes fixed resolutely on an ostrich that was daintily picking her way through the bush on the other side of the jeep. We watched these two enthusiastic males go at it for around thirty minutes (actually only thirty seconds but surely the longest half minute in the history of time) and no one said a word. I couldn't look at the boys. You didn't see this in *The Lion King*.

'In fact, many animals display this kind of behaviour. Dry humping is an important ritual in adolescent behaviour.'

Freddie, who may not have been watching but who was definitely listening, pointed at Josh and Ben. 'You two are adolescents.' Ben reached over and thumped Freddie hard on the arm.

I mentally pleaded with Priscus not to venture on to the alternative of 'dry humping', which I guessed was 'wet humping'. It was a bizarre moment of trying to decide what would be worse to hear, while stuck in the middle of the African savannah with my three sons.

'Yes, of all the behaviour you'll see here. Dry humping is probably the most common.'

There was something about this term that made my toes curl. It was just so damn graphic.

'So, shall we move on?' I suggested brightly.

'You okay, Mum?' asked Josh. Soon to be sixteen, he had a new man-of-the-world aura about him that I blamed on playing rugby.

'Fine,' I lied, busying myself with cleaning the lens on my camera.

I'm not normally the kind of mum who shies away from this kind of stuff. When Josh was around the age of nine he came home from school and announced that he knew how babies got into mummy's tummies (he was at a Church of England School so I guessed correctly that God might be involved), which triggered a whole conversation about reproduction (including the mechanics of sexual acts) that I hadn't anticipated happening quite so soon. 'Any questions?' I'd asked him after delivering the basics of copulation. 'Can I have a piece of chocolate cake?' he asked. Handily for me, Ben informed me just an hour later that he also knew all there was to know on the subject so wouldn't need me to go over it again. Oh, the joy of having an older brother who was unable to keep his mouth shut. And I was pretty sure that Freddie, with two older brothers, would have it all worked out by now too. After all, he was a typical third child, pretty much dragging himself up. And if he hadn't quite got the full picture, well this holiday was certain to fill in the rest.

When we got back to the lodge, the boys trooped off to find Lemuani, keen to view the footage that the camera traps had recorded the night before.

'Not like you to be so prudish,' Neil teased me once we were alone. 'I saw you squirming out there.'

'Not every day we watch the equivalent of monkey porn with the boys is it?' I snapped back.

'Dry humping,' Neil corrected me. If I never hear that term again it will be too soon.

'Not so squeamish about the zebra and her sperm-expelling skills, though,' commented Neil

'Handy trick to have up your . . .' Thankfully, I saw the joke quicker than Neil. At times, it wasn't easy being the only female in the family.

When the boys came back they were excited. 'There was an enormous male lion sniffing around the perimeter fence last night,' Ben told us. 'Just ten metres from our room.'

What a happy, fortuitous coincidence then that I'd opted for a bush dinner on that very evening.

'You comfortable with all of this?' I asked Neil later, once we were sitting in front of a campfire, around 100 metres away from the safety of the lodge.

'Not really. You saw that lion. It was a beast.'

'Are you winding me up?'

He nodded towards a National Park guard who was standing nearby. 'At least we've got him,' he said.

The guard had a gun. A detail that hadn't gone unnoticed by the boys. This was quite a step up from a spear and a knife.

Having dinner out in the wild had seemed like such a good idea, but the reality of feeling quite this exposed to predators was unnerving. I watched the sun turn from gold to crimson and dusk fall, in between glancing over my shoulder. A barbecue was being stoked behind us and a chef was busy seasoning steak ready to cook for our supper. Steak! I mean, what could possibly go wrong? I wondered how long before these delicious meaty aromas would attract wild guests.

'But at least we're eating up there,' Neil said, pointing to where a table had been positioned on a wide flat rock, the very kind of rock that Priscus had told us that lions loved to hang out on.

'If a lion came from behind the guard, we'd all be toast,' I said.

'Not really toast. More like chops,' Neil corrected me.

'Let's go to the rock, shall we?' I suggested, jumping up quickly.

Turned out that this was one of my better ideas, as a few minutes later a pack of hyena gate crashed our party. We listened to them cackling below and when I caught sight of their hunched-up bodies, lolloping past in that shifty, off-kilter way of theirs I felt the hair on my arms bristle. Neil remained unfazed, making jokes about how Freddie would make a nice starter, Josh a tasty main course and Ben a lovely desert. The guard hissed at the hyenas and stomped his feet, and they ran off chuckling.

'Thank God,' I said. 'Let's just eat fast and get back to the lodge.'

'Oh, Mum, we're fine,' said Josh.

'What's that noise?' asked Freddie, as we picked up our knives and forks to tuck in.

The hyenas hadn't got very far. Nearby, a pair began to grunt their way through a main course of quite a different kind.

'Delicious steak,' I commented, while Neil and the boys' laughter, like an elephant's trumpet, could most likely be heard 50 kilometres away.

13

Japan

A Lesson in Social Etiquette

Josh age sixteen, Ben fifteen and Freddie eleven

10 April 2017,
Kyoto, Japan

The boys and Neil aren't completely sure about my choice of accommodation – a traditional Japanese 'ryokan' inn – but I love the simplicity of it all. Our communal bedroom ('What? All of us in one room? No way!') has a tatami mat floor and bedrolls to be unfurled at bedtime. A low table and chairs without legs are the only other pieces of furniture. I haven't yet broken it to Neil how much we're paying to stay here (Japan is a clear winner in the most expensive trip ever stakes). There's not even a mini bar full of Sake to help him forget, once I do.

Ignoring my family's general negative vibe (perhaps the ryokan *has got its feng shui wrong?) I am determined to embrace the experience. We've each been given a* yukata *(cotton kimono), along with flip-flop style slippers and a strange pair of socks, with a separate sleeve for your big toe*

– cleverly designed to wear with thonged footwear. Neil and Freddie are the only ones joining me in turning Japanese, with Freddie enthusiastically pulling on what he is calling his 'glove socks'.

Typical of all ryokan *are the communal baths, fed by natural thermal springs. At the entrance to them, we were handed two essential bits of kit by the attendant – a flannel-sized towel and some body wash. 'No shaving in pool, please' he informed Neil.*

'Er, weird. Who would shave in a pool?' asked Ben.

'And this is the stupidest towel ever. How can you dry yourself with that?' Josh wanted to know.

I knew the score. This was a special, rather intimate little towel, used for covering one's privates. Thank God, the pool wasn't unisex.

'Must go naked please,' the attendant stated bluntly.

'Say what?' asked Josh.

Ben did a swift about turn. 'No way. See you back at the room,' he said, stomping back towards the stairs.

~~~

It was cherry blossom season in Tokyo and the city was at its most beautiful with over a thousand trees in Shinjuku Gyoen Park heavy with pink and white blooms and packed with locals enjoying their annual *sakuru* (cherry blossom knees up). For me, this was the cherry on the cake in terms of travel writing as

Japan had always captured my imagination, from my obsession with the Eighties pop band Japan, to the excitement over my very first revolving food belt experience in 1997, to gazing at the *netsuke* (sixteenth-century decorative miniature belt charms) in the Victoria & Albert Museum, London. My latest fascination was with its wacky pop culture, which I was hoping my teenage sons might help me navigate.

'What the hell?' queried Josh, as a parade of Young Japanese women, dressed as Little Bo Beep, Queen Victoria and what looked like Lady Gaga, queued up to pose by the cherry trees to have their photos taken. It looked like Josh may need a little help in understanding it all, too.

'Perhaps better to expect the unexpected?' I suggested, as a Dutch milkmaid clopped past us in clogs.

With the Japanese obsession with cosplay (the practise of dressing up as a character from a film, book or video game) and with anime (the Japanese hand-drawn or computer-generated cartoon, which also inspires role play), it was common to see people dressed in fanciful outfits. Just about as culturally different as it gets, it was best not to view Japan simply as weird, but rather weird and wonderful.

Of all the places we'd taken the boys to, the potential for cultural faux pas were the greatest here. I'd researched rules of social etiquette before arriving and had given the boys a briefing. For example, they were never to sit with their legs open but instead slant their legs and feet to the right (straight towards a person was an omen of death); they were never to eat or drink in public while walking; never to leave their chopsticks pointing upwards in their bowls and never to use their chopsticks to point or poke, and (oh, how I loved this one) being a picky eater was considered very disrespectful. Raising your voice in public was judged the height of bad manners, and as I was accustomed to being able to holler loudly to stop the boys from doing anything that I considered unacceptable or dangerous – as in 'Get down from there!' 'Stop fighting!' 'Watch what you're doing, for Christ's sake!' to the most common, a simple anguished, 'Stooooooooooop!' – this would clearly be my greatest challenge. You'd think perhaps that by now, with sons aged sixteen, fifteen and eleven, my days of doing this would be over, but you'd be wrong. If anything, there was more scope for danger.

Tokyo, a vast metropolis that provides little information in English, is daunting, even for the most seasoned of travellers and we were no exception. There'd already been some confusion over dinner, as what had arrived had born no resemblance to what we'd ordered, and the ensuing kerfuffle was straight out of the film *Lost in Translation*.[23] We'd all got the

---

[23] *Lost in Translation* is the 2003 comedy-drama film set in Tokyo, which explores the theme of cultural displacement.

jet-lag giggles – a kind of laughing/crying hysteria – while dining on a weird mixture of glutinous jelly dessert and curried rice with seaweed that had given me indigestion all night. In line with the Japanese etiquette of not making a fuss in public or being a picky eater, I'd found myself hissing, 'Just bloody well eat it' at the boys.

To ease our exploration of the city, I'd booked a guide. The brief I'd given her was that we'd like to experience anything that was typically Japanese but that we, as foreign visitors, might find unusual (although made sure to mention the boy's ages – the fetish scene in this city is off the kinky radar). Yuna got off to a flying start. Where better to suggest meeting a family with three boys than Philippe Stark's curiously shaped 300-tonne *Flamme d'Or* (golden flame), which is fondly and widely referred to as the 'golden turd'.

'What someone actually designed it to look like a poo?' Freddie asked.

'Like fire,' Yuna explained.

'A poo on fire?'

'Haha! Very funny,' said Yuna. Thus, our accidental three-hour family stand-up comedy routine began.

Apart from bandying words such as 'turd' around, Yuna's manners were impeccable in that bowing frequently, softly spoken reverential way that Japanese people have about them. The fact she appeared to find us amusing company, tittering behind her hand every time one of us asked a question, was fine. Soon we were all laughing, although no one was quite sure why.

'First stop very close by,' she told us, as we followed her through the Akihabara district, known as 'geek city'. Packed full of *gachapon* arcades (the simple, though somewhat addictive pleasure of putting a hundred yen into a slot and seeing what pops out encased in a plastic capsule), we made a few unscheduled stops for Freddie to start his collection of Japanese memorabilia (utter plastic tat). We also made a small detour to a street that sold only plastic food, manufactured to display outside restaurants and created in such authentic detail that it was hard to believe that they were fake, particularly the frothy pints of beer, plates of sushi and bowls of noodles. We settled on buying a bowl of ice cream that looked like it had been freshly scooped from the tub. 'We'll fool Grandpa with that one,' plotted Freddie.

Easily distracted by everything around us, Yuna hurried us on until we came to a halt outside a place with smoky grey glass windows and a cartoony logo of a French maid above the door.

'Nice for teenage boys,' she said, smiling at Josh while explaining where we were, which was, in fact, a Maid Café. I knew of these coffee bars, where waitresses dressed as French maids made boys without girlfriends (in a city where women outnumber men) feel special. 'Hello, Darling. I've missed you,' was the usual sort of greeting. Word was that it was innocent stuff, with no solicitations beyond the request for coffee, but it still felt very wrong. As a mother of three sons, I was vigilant (bordering on obsessive) in stamping out sexism. This sexy, coquettish stereotype of a female maid, there to please male customers (even if it was only in serving coffee), made me very uncomfortable.

'Oh, I don't think this is for us,' I said.

'Nice for lonely boys,' said Yuna, looking again at Josh. I heard Freddie and Ben snort with laughter.

'There is no way I'm going in there,' Josh hissed.

With my journalist's hat on, I had to admit to being curious (more than anything so that I could write with first-hand experience about how awful these cafes were) but asking my husband and sons to wait outside while I investigated wasn't an option. I could tell that my 'all-in-the-name-of-research' card that I sometimes played would be ripped into tiny shreds should I try to use it here as they were all clearly squirming with embarrassment at even being in the vicinity of the place.

'I think we'll move on,' I told Yuna, who looked devastated at having made a bad call in bringing us to somewhere we were clearly feeling very uncomfortable about.

'So sorry,' she said.

'It's fine,' I told her.

'So, so, sorry.'

'Thank you, but really it's not a problem. Perhaps there's somewhere else we could have a drink?'

'Again, so sorry.'

Apologizing in Japan is more than simply saying you are sorry. It's about reflecting on what went wrong and the effect your actions has had on another person, and Yuna wasn't letting herself off the hook easily.

'Yes. I can see you are, but really it's my fault. I'm sorry. I should have been clearer in telling you the kind of things we'd like to see.'

'Oh no. Please. It is me who must say sorry.'

I hesitated then. How long could this go on? I could see that Neil and the boys were fascinated to see how this game of apology ping pong would end.

'Your apology is of course accepted. Please also accept mine,' I said, trying to sound a touch firmer.

Yuna seemed happy with this and bowed her head in what I hoped would end the game at deuce.

'You like cats?' she asked, taking the conversation in a sudden surprising direction.

'Didn't see that question coming,' remarked Neil.

We nodded and Yuna tittered behind her hand again.

'Not like you,' commented Neil, once Yuna was out of ear shot.

'What?'

'To be quite so sorry.'

'It's the new polite Japanese Kate,' I told him.

If Akihabara is for earnest looking teenage boys searching for components to make God knows what (perhaps a girlfriend?) in their bedrooms, then Harajuku is a Mecca for teenage girls.

'And what on earth are we doing here?' asked Ben, as we stared down Takeshita (queue the sniggers) street, at a sea of girls and shop after shop of make-up stores and teen fashion emporiums.

'Something to do with cats,' I whispered, just as a passing girl with blue hair wearing a zebra onesie squealed '*Aww, kawaii*' (aww, cute) at Freddie. Poor Freddie. It was a sentence that was to be on repeat to him from his not-so-kawaii brothers for the next two weeks.

The concept of a cat café (along with rabbit and also owl cafés) began in Tokyo, where high-rise living is not conducive to the keeping of pets. The Cat Café is exactly what it says on the tin, somewhere you can go to get a drink and stroke a cat, and as the cats aren't contractually obliged to wear French knickers or a lacy cap we were all a lot more comfortable with the idea. As usual in Japan, there was a rule: the cat must come to you. Predictably, the challenge to lure a cat on to your knee got competitive, with Freddie, the most successful animal-whisperer in the family, winning every time. 'You're hogging all the cats,' was the rather unusual gripe from the other two.

We parted ways with Yuna after more bowing, a few more apologies, and a lunch stop at a small sushi restaurant where the conversation had gone something like this:

Me: 'Chopsticks flat down!'

Freddie: 'Sorry!'

Ben: 'I'm not eating that.'

Me: 'Picky eater, very rude!'

Me: 'Stop waving your chopsticks about!'

Freddie: 'Sorry!'

Josh: 'That tastes horrible!'

Me: 'Picky eater, very rude!'

Me: 'Not in the bowl!!!! Place them on the chopstick rest!'

Freddie: 'Sorry!'

Ben: 'What even is that?'

Me: 'Picky eater, very rude!'

Me: 'Stop pointing with your chopsticks!!'

Josh, Ben and Freddie: 'Mum, stop shouting! You're embarrassing us.'

Yuna, I think, was glad to see the back of us. Without her, we all looked to Neil to take over as navigator. Like a superhero, he strode towards the nearest subway and stared at a map on the wall, so complicated that just looking at it gave me palpitations. To put things into perspective, there are 882 interconnected rail stations in the Tokyo metropolis, and 282 of those are subway stations, and an average of 7.6 million passengers travel between them each day, easily, efficiently and quietly (if you don't count us). We'd promised Freddie that we would find him a Pokémon shop and had located one in the basement of a large shopping mall, only three stops away with one change. The etiquette on the subway was much like it is on the tube in London, namely eye contact is not welcome and personal space is appreciated, but there's one big difference – no one makes a sound (not even if they were travelling with friends or family) and no one ever speaks on a cell phone. In this over-populated mega-city, invading people's ear space was the worst you could do.

'Oh, shit! That was our stop,' Neil said, jumping up just as the train was gliding away from the platform.

'Oh, for God's sake. You had one job,' I said, sounding every bit as cross as I was.

'Everyone's looking at you,' Freddie told us.

It was true. Not only had we spoken, but in raised tones.

'So, what now?' I hissed at Neil.

Neil scanned the subway map in the guidebook.

'I think stay on,' he whispered at last. 'And get off in two stops time.'

'Are you sure?'

'No, Kate. I'm not sure, but you have a go at getting us there if you think you can do better.'

An hour later, having disembarked to discover there was no connection to our desired destination, and trying another (wrong) subway line, we were back on the first line going towards the stop we'd missed an hour ago. By now, no one was talking to each other, so silence wasn't a problem. Nice, polite, Japanese Kate had long since pissed off and impatient, quick-tempered Kate was back – the one Neil had married.

At the Pokémon store, Josh and Ben were grumbling. If we'd followed their itinerary we'd be in the BAPE STORE browsing exorbitantly priced Japanese street wear that was, according to them, difficult to buy outside of Japan.

'We've wasted two hours following Dad about underground and now we're stuck here,' moaned Ben.

'And there was a drop this morning,' huffed Josh.[24]

They were at an age now that a request to strike out to explore without us wouldn't be unusual, but not even they were brave enough to suggest going it alone here.

---

[24] A drop is a limited release of merchandise from a streetwear brand.

As always, I was conscious that Freddie, the youngest and most easy-going of my three sons, often tagged along on whatever the rest of us had decided to do.

'You enjoy yourself,' I told him.

After an hour in the shop he'd settled on some socks and a tin of cards. 'Is that all?' the other two wanted to know, when Freddie showed his brothers what he'd chosen. 'All that time and effort to get here and you're buying socks and some cards?' Fast forward four years to Freddie selling a card in that pack for ninety pounds on eBay (kerching!) and Josh and Ben weren't quite so scathing.

There are certain things in Tokyo that as a tourist you're required to tick off and Shibuya Crossing, the world's busiest road crossing, is one. Looking down on it, from a bird's eye perch in Starbucks, I felt my heart rate speed up. To the untrained eye, it looked like utter chaos and, to make matters worse, Josh had insisted that we wait until 6p.m. to experience what is known as 'the scramble', when office workers spill from their workplaces in the sky.

'Oh God, let's at least try to stick together,' I told them. 'Walk normally for once. You know, like in a straight line.'

Josh and Ben rolled their eyes simultaneously. 'We're not five,' Ben said. 'We can cross a road.'

'But this isn't any old road. This is the busiest road

in the world!' I think my voice may have gone a little shrill as several people looked our way.

'It's you who needs to act normal,' Josh told me.

As 6 o'clock loomed, I felt like I was launching myself from the trenches as I teetered on the curb, marvelling at the expert way the Tokyoites manage to duck and weave, never bumping into each other. If we walked in a straight line and let the locals do the dodging, we stood a fighting chance of not being bowled over. I kept a firm grip on Freddie's hand, wished the rest of them good luck and walked resolutely in a straight line. There were several alarming moments when I thought I might collide head on with a Tokyoite office worker hurrying home to his sushi supper, but resisting the urge to swerve, my tactics paid off. It was quite the adrenaline rush. Under a huge neon sign advertising something called the Robot Restaurant (bento box-style dining with a robot show), Freddie, Josh, Ben and I waited for Neil.

'That looks good,' Freddie said, pointing up to the sign where a neon green transformer-type man was playing electric guitar with a tiger roaring behind him.

'Does it?' I said, waving frantically in an attempt to get Neil's attention. He was still caught up in the throng, and had veered way off course, engaged in what looked like a game of human bumper cars.

'Come on, Mum. Be the fun parent,' Josh suggested, which was what the boys said when they were trying to guilt-trip me into agreeing to something that they guessed (always correctly) I'd be reluctant to do.

'Let's do it!' I said, before Neil arrived and won the fun parent title (or rather, damn him, retained it).

So as not to be a kill joy, once inside I resisted in pointing out that this wasn't in fact a restaurant. They had lied on the poster. Sure, you could buy bento boxes, but no one did, because there weren't any tables to put them on. Instead we were rammed into tight rows, conned into buying expensive fizzy pop and beer and advised that during the performance that it was probably better not to leave your seat (which triggered a bout of claustrophobia not at all conducive to the behaviour of a 'fun mum'). Just as I was contemplating making a scene by climbing over the three rows of people in front of us and legging it the hell out of there, the lights went off and I was trapped. For 90 minutes, an unrelenting parade of UTTER MADNESS passed us by – giant sparkling fish accompanied by dancing mermaids, phoenix-like birds escorted by drum-majors, dragons that breathed neon smoke, unicorns ridden by clowns, Zulu warriors cavorting on Cadillacs, punk rockers stomping on monster trucks. None of it made sense. Add techno-music, flashing lights, strobes, a giant t-rex and a robot boxing match and the warning that it was not suitable for those suffering from epilepsy, should have come with a broader caution – that for anyone over the age of thirty your eyeballs would most likely melt and your head would explode. If the boys heard me mutter, 'Please make it stop,' they didn't let on, unanimously agreeing when it had at last (thank you, God) ended that it was the best thing they'd ever seen.

'No tiger though,' commented Freddie.

'No dinner, no tiger and possibly the most insane thing you'll ever see in your life,' I told them.

Back at the hotel, I crashed on the bed face down. Neil had gone for a beer in the hotel bar, but I didn't even have the energy to lift a glass to my lips. It was mind-blowing to think that we'd only been in Tokyo for a day. Five minutes later, there was a knock on the door.

'Can't move,' I yelled.

'It's me,' shouted Freddie.

When I opened the door, I saw that Freddie had the kind of shining eyes he got when there was trouble that he wasn't involved in. 'You've got to come,' he said.

'Really?' I asked. That involved walking a full five metres and I wasn't sure I could.

'Josh and Ben are fighting.'

There'd been a scrap. A battle over the only can of lemonade in the minibar and who should have it. Josh had hit Ben, and Ben had thumped him back. In the tussle the lemonade had been spilled over the bed and carpet. I'd only been in the room a nanosecond when Josh stepped forward.

'Look, Mum. Before you say anything. I'm really sorry.'

'Me too. Sorry,' Ben echoed.

I opened my mouth to speak, but Josh hadn't finished. 'And I'm really sorry that I hit you, Ben.'

'And I'm sorry I hit you back,' Ben answered.

'Hmph,' was all I could manage.

Getting these two to apologize to each other had always been a struggle. After all, the word 'sorry' really has no meaning if it is followed by a swift 'but I don't really mean it' whispered under your breath, which was often the case following a fall out.

I narrowed my eyes. 'Are you though? Really sorry?'

'We really are,' said Josh.

'We really, really, are,' agreed Ben.

Did I see Ben smirk at Josh? Perhaps I did, but I'd give them the benefit of the doubt this time. Too knackered to coax them into reflecting on their actions, that was something we could work on (or not. I had to be realistic) after a good night's sleep. Whatever their motives, I could see that Japan was having a positive effect on these two, at times stubborn/competitive/argumentative/perfectly normal brothers, even if the apology was only being used to get them out of trouble quickly with their, at times, embarrassing/annoying/short-tempered/perfectly normal mum.

14

Laos
A Lesson in Slowing Down
& Letting Go

Josh age seventeen, Ben sixteen and Freddie twelve

2 August 2018,
Luang Prabang, Laos

In the tiny private garden of my hotel room there is a bathtub on a platform, nestled among ferns. Late afternoon, while Neil and the boys swam at the pool, I ran a bubble bath and soaked for an hour, while watching the snails make slow progress across the wall – a creature so perfectly suited to the pace of life here.

The Wi-Fi signal is hit and miss, something that the boys aren't happy about, but I'm hoping that means more time spent playing cards and chatting rather than staring at our phones.

The hotel was a prison until 2006, and the tuk-tuk drivers are a hoot, refusing to recognize it as anything else. I gave the name of the hotel to one and he immediately scratched his head.

'Not know it,' he said.

'Are you sure you don't know it?' I coaxed. 'It's just by the bomb museum.'

'Ohhhhh! Prison hotel!'

It's obvious that persisting in calling it by its fancy new name won't get us home.

I'm guessing that our rooms, lovely though they are with beds canopied in mosquito nets, hard-carved teak furniture and Lao silks on the walls, are actually converted cells. The boys have been cracking jokes about it all day. 'What you in for?' I heard Ben inquiring of Freddie.

'Drugs but I was framed,' was Freddie's reply.

I'd like to say that without knowing of its history (the hotel describes it as a former colonial mansion, which it might well have been before it was a gaol), you'd never guess, but the watchtowers, though prettily white-washed and climbing with flowers and vines, do rather give the game away. We climbed the stairs of one for views over the hotel's gardens.

'Escapee, six o'clock,' whispered Freddie, referring to one of the hotel's two resident pet rabbits that was happily hopping alongside a lily pond.

An old Mercedes sits in the courtyard with a sign that reads 'Welcome Home' where its number plate once hung. I'm not sure that former prisoners would be reassured by such a friendly greeting, but for those of us who are humouring the new owner's story, it's a nice touch.

Time spent in Luang Prabang, once the Royal capital of Laos and now a UNESCO World Heritage Site, gave us the opportunity to practise slow travel, which was all about connecting with the local people and culture in an unhurried, positive way. We arrived there from ten days in Bali, where you had to work hard to find places that had not fallen foul of over-tourism. The roads were grid-locked, the most scenic spots and temples packed full of tourists and much of the development around Kuta was brash, ruining this once paradisiacal beach resort. Arriving in sleepy Luang Prabang felt wonderful. It wasn't quite an undiscovered nirvana, but it came pretty close, with lodgings in Indochinese and French colonial villas set in lush gardens, restaurants lit by lantern-strung Tamarind trees and gleaming temples, home to saffron-robed monks. I wanted the boys to get a feel for what it had been like for us backpacking around South-east Asia in the mid nineties without a phone to keep us in constant contact with our family and friends (cue the eye-rolling agony at that awful thought). I was yet to break it to them that I wanted all of us to be off comms for much of our stay. The Wi-Fi signal was in their words 'rubbish' anyway and had already caused a couple of minor melt downs. 'I can't do my streaks,' had been the most recent lament. Apparently Josh was on a three-hundred-and-something-day 'streak' on Snapchat with one of his friends, which he could not, under any circumstance, break.[25] It had been a struggle to work up any sympathy.

Josh, Ben and Freddie were used to the packed itineraries I normally presented them with, so it took a bit of time to get used to the idea that here we would be taking it much slower.

'So, basically we're here to do nothing?' asked Josh, who was a keen surfer and had taken every chance to be out on a board in Bali.

'I'm happy just to stay by the pool,' said Ben, which wasn't quite what I had in mind either.

'Well, not exactly nothing. We'll take a river trip on the Mekong. We'll cycle around town to visit the temples.'

'And the rest of the week?'

'Oh, we'll just go with the flow, shall we?' I answered, resisting the urge to run back to the room and rummage through my guidebook for ideas, or worse still turn to Google.

As the hotel provided free bikes, this seemed like a good way to explore – the roads were flat, the locals (all Buddhist) gentle and polite, and Luang Prabang is a small city and easily navigated. We pedalled away with me screaming instructions for

---

[25] Streaks count how many consecutive days two people have been sending Snaps to each other on Snapchat.

the boys to keep both hands on their handlebars, to wait at junctions and, if in doubt, to give way. Why we never had these discussions before setting off was anyone's guess. As usual, not one of them was paying any attention, too keen to take the lead, and within minutes I saw Josh take not one hand but both off the handlebars.

'Mum! Come on!'

'Where's Mum gone?'

'What is she looking at now?'

'Have we lost her again?'

Despite their loud complaints and frequent calls for me to 'get a move on', I ignored them all and continued to travel at my own pace. The Lao people joke that the abbreviation 'PDR', as in (The Lao) People's Democratic Republic, really stands for 'Please Don't Rush', and I thought this might be a good time to mention it. We were cycling along the banks of the Mekong River, which, thanks to its muddy silt banks, flows with what looks like chocolate milk.

'Keep up, Mum,' shouted Josh.

Ben had already cycled on, fed up with my insistence in stopping every few pedals to gawp at something or snap a photo.

'We're not in a hurry are we?' I shouted back.

'No, but if we go any slower, we will actually be travelling backwards,' he groaned.

'But there's so much to see,' I argued. Home to over thirty-three gilded wats (Buddhist temples), where monks lived, worked and studied, the joy here was in the minutiae of everyday life. I stopped to observe a young novice monk being tutored in verse by an older teenage monk in the grounds of one.

'Did you see those young monks back there? One was about your age, Freddie? Did you see him?' I asked, when at last I'd caught them up.

They hadn't, of course. They'd been too busy doing wheelies.

'You need to look around,' I told them. 'Don't look down. Look up. You're missing so much.'

'Stop worrying about what the boys are doing and just enjoy yourself,' Neil told me, as we pulled up to a café called Saffron, which, according to the sign outside, supported northern Lao coffee-growing hill tribes.

'You can't tell us where to look,' said Ben. 'That's ridiculous. God, you're even trying to control our eyes now.'

If there was a button to push in me, this was it. He was making me sound like a helicopter mum, forever hovering over them and interfering, and this irritated me for two reasons – one, the comment about controlling their eyes was actually pretty funny, and two, there might have been just the tiniest grain of truth in it. These days, aged seventeen and sixteen, Josh and Ben, were only too quick to pick Neil and me up on parenting crimes, and helicopter parenting was about as serious as it could get. At times I hankered after those early years of travelling, when they weren't nearly as opinionated, but most of the time I loved watching their journey from childhood to manhood, and all that it entailed – the challenges to our decisions, the debate of opinion, the frequent refusals to do as they were told. Frustrating as this transition could be, we were really proud of the young men they were becoming.

'I just don't want you to miss out, is all,' I told them.

'Well, that's our problem,' said Josh. 'Not yours.'

Josh had just sat his A levels, Ben was starting his and Freddie had finished his first year at senior school. Taking a couple of weeks away from their busy lifestyles of sport and socializing, schoolwork and PlayStation addictions (Freddie was the worst) would do them the world of good. What I couldn't control, however, was how much they would embrace the opportunity. Josh was right. That was their problem.

We sat on Saffron's terrace and looked down on the Mekong. Below us, captains of long-tail boats touted gently for business and fishing boats and small ferries floated by. Josh and Ben pulled out their phones (our agreement to leave them at the hotel either already forgotten or most likely ignored) and stared at the screens.

'I thought . . .' I began.

'Huh?' asked Ben, looking up briefly.

'Nothing,' I said.

We cycled on to Mount Phou Si, where we left our bikes to climb 329 steps to the summit, past statues of golden Buddhas, to where a gilded stupa crowns the hill. At the base of the stupa sat a lady selling birds, which she'd trapped in tiny pink and yellow rattan woven cages. I'd seen this practice before in South-east Asia and I didn't like it. The belief was that releasing the birds would bring you good luck. The cages were small, the birds desperate to take flight, and their anxiety palpable as the tiny cages twitched with the birds attempts to flap their wings. Nearby, a cat was resting, clearly enjoying the show.

'How much?' I heard Josh ask the bird lady.

She held up four fingers to indicate four thousand Lao kip. The equivalent of thirty pence.

'We could set them all free,' suggested Freddie.

Amid the cluster of swinging cages held up for our perusal, one bird was lying motionless, already dead. 'We shouldn't buy them,' I told the boys turning away.

'Oh, come on, Mum,' Ben said. 'You can see they're unhappy in those tiny cages.'

'But once those are gone, she'll just trap more,' I told them.

'We should at least save these,' begged Freddie.

I took another look. One peeped pathetically at me. It was one of those moments I'd experienced so often while travelling. Do I support a business I don't agree with (even for good reasons), or walk away and let the birds remain captive? Business was slow, and it was likely more would die.

'And it'll be lovely watching them fly away,' added Freddie.

'Your decision. I've told you what I think,' I said.

The boys bought one each. Their intentions were good, only for the welfare of the birds, and possibly a little to prove to me that they were their own people now, free as a bird (annoyingly, the symbolism wasn't lost on me) to make decisions without needing my approval.

Freddie glanced at me nervously. Of the three, he was the most reluctant rebel. 'Alright, Mum?' he asked as we watched them fly away. I nodded. He was right. It was lovely to see them fly away.

Back at the hotel swimming pool, I watched Ben pick up his book. 'Think I'll just stay here and relax,' he said.

My suggestion had been to climb to the top of a nearby hill to watch the sunset, but the boys were having none of it.

'I'm staying too,' said Freddie, jumping into the pool.

'We'll be fine here. You go,' encouraged Josh, rubbing on more sun cream and settling back on his sun lounger. Neil also looked content, having just ordered a beer.

'Let's do it another day,' he suggested.

I reached for my book, batting away my FOMOOS (fear of missing out on sunsets). It was a family joke that we often failed to see them by just a few minutes, which triggered my usual rant of: 'If you'd just come when I told you instead of messing around we would have seen it. I bet it was lovely and now we've missed it AGAIN!' In Bali, the sunsets had been incredible, if you didn't mind viewing them through a forest of arms belonging to the fifty million other tourists at the cliff top temple/beach/rice paddy all filming it on their phones. For once, though, I decided not to argue. This slower way of travelling was doing us all good. Here, the sunsets belonged to us, and there'd be another one tomorrow.

Sticking to my travelling habit, I booked the services of a local guide. In Luang Prabang, this was Kharta, an open-book of a chap, who within minutes had shared his life story, impressing us all with his claim that he'd been born in a cave on the Ho Chi Minh Trail – the system of mountain and jungle paths that the North Vietnamese used to infiltrate troops and to transport supplies into South Vietnam during the Vietnam War. 'My mother couldn't bear to leave my father's side,' he explained.

'Would you have followed Dad?' Freddie wanted to know.

A dim recollection of using morning sickness as an excuse to wriggle out of accompanying Neil on a works trip to Paris flooded back. If I couldn't work up the enthusiasm to stay in a swanky five-star hotel on the Rue du Faubourg while pregnant, I thought it unlikely that a cave on the Ho Chi Minh trail would tempt me.

'Of course, I would,' I fibbed.

Kharta was taking us up-river by boat to visit a local village and then on to the Pak Ou Caves, home to hundreds of Buddha statues. Just out of Luang Prabang, we passed fisherman casting broad nets for bass, and a gilded wat flanked by thick vegetation where monks, who were constructing a new pontoon, had stopped for a tea break.

Kharta smiled when he saw them. 'Ah, they are putting into practice the Lao belief that too much work is bad for you.'

'See, Mum. I told you,' commented Josh, referring to the work he'd been cajoled into doing for his recent A levels (by his helicopter mum. Whatever).

'Hmmmm. We'll see,' I said. He hadn't had the results yet.

'There's a saying that the Vietnamese plant the rice and the Lao watch it grow.'

'That makes the Lao people sound a bit lazy, doesn't it?' I asked.

'It doesn't mean we're lazy. It means that we know how to be still,' Kharta said. 'Not many people are good at that. Doing nothing takes practice.'

I knew what Kharta meant and it was something I was trying to address in my own life. The need to always be doing something (as in trekking up a hill to watch the sun set) was a hard habit to kick.

We disembarked at Ban Xanghai, a whisky-producing village, following Kharta into a ramshackle shop, which looked like a cross between a medieval apothecary and an off-licence. We were here to learn about the distillery process but it was hard to concentrate; all too distracted by the snakes and scorpions that were pickled in bottles on the shelves.

'I see you're curious about these,' Kharta said finally, holding up a bottle that had a scorpion floating within. 'Lao people believe that if you drink this, you'll be immune to the creature's venom.'

'What about that one?' Ben wanted to know, pointing to a much larger canister, containing a somewhat wrinkled item.

'Elephant penis,' Kharta told him, opening the jar and taking a sniff. 'Good for sexual potency. You want to try some, Dad?'

'I think three boys are enough,' Neil said, winking at me.

'Oh, God. Kill me now,' I heard Ben mutter. Josh had once suggested that we'd only ever 'done it' three times, in order to produce them, and it was obvious that the suggestion that we might still be 'at it' was nothing less than horrific.

'What's that?' asked Josh, pointing to a pipe tucked away at the back of the shelf.

'Opium pipe,' Kharta told him.

'How much is it?' Josh asked. Dear Lord, we'd come a long way from buying Spiderman sandals in the markets of Bangkok. 'Don't worry. It's just for decoration. I won't actually smoke anything in it,' he told me, as I watched him hand over his money.

At the Pak Ou Caves, Freddie hesitated. 'There won't be cockroaches in the cave will there?' he wanted to know. Cockroaches were his nemesis, capable of turning my normally plucky youngest son into a gibbering wreck.

It used to irritate me that the boys expected me to have the answers to questions like this, but as the frequency that the boys looked to me as 'knower of all things' was dwindling, I was now hanging on to these moments. 'No, of course not,' I said, winging it as usual.

Damp, dark and wonderfully creepy, we scrabbled up slippery moss-covered stone steps to the dingy centre of the main cave where thousands of Buddha effigies, the oldest of which dated from the fourteenth century, sat watching us.

'Imagine being here at night,' Freddie whispered, while we watched Neil freak out after a drop of water landed on his head.

'You do bring us to weird places,' Ben said. Well, I couldn't argue with that observation. Although, it did seem a little odd that, after all these years, Ben was only just making it. From the 'bat-poo' caves to Borneo to Pak Ou cave of Laos, there was certainly a theme.

'Never boring though, is it?' I answered. 'And by the way, I'm having a day off tomorrow, so you lot can plan the day.'

It was always fun for Neil and me to discover what the boys had in store for us, as often it was an activity that I'd never have thought of. Which was how we found ourselves, the following day, bouncing in the back of a tuk-tuk truck. The night before, I'd had one too many Herbal Remedy of Uncle Ho (a lethal but delicious cocktail made with Lao rice whisky, banana liquor, basil and coconut juice), and I was feeling a little delicate.

'It's the most fun way to get around,' Freddie explained, when I groaned at the sight of our transport.

Well, that wasn't true. Clearly more comfortable for short journeys, our driver, the Lewis Hamilton of the Formula One tuk-tuk driving world, was providing us with a white-knuckle ride, throwing us around like peas in a tin.

'He needs to slow down,' I shouted.

Neil rapped on the glass that separated us from our chauffeur (in his nice safe, closed compartment). Alarmingly, he took his eyes off the road to watch Neil perform international sign language for slow down – by moving his hand slowly up and down. The driver nodded, grinned and then put his foot down to go faster.

Over wooden bridges and through farmland of fluorescent green rice paddies, with my bottom rarely resting on the seat, I caught glimpses of idyllic countryside life flashing past us, where buffalos wallowed in mud-slicks and kids crouched by river tributaries to fish.

'This is not slow travel,' I heard myself screaming at one point. The boys, however, were having a ball, whooping at every jolt and judder in the potholed road until we reached our destination, the Kuang Si waterfalls.

'Oh, beautiful,' I gasped, gazing at three-tiers of frothy falls cascading from a height of 60 metres into menthol-blue pools, which were also clearly hazardous. 'No Swimming' signs had been placed next to two of the pools, but at others, that looked equally, if not more dangerous, there were none. This haphazard way of dealing with safety issues was so typical of South-east Asia. In the end you had to evaluate the risks yourself, and Josh, Ben and Freddie, having given the sign the most cursory of glances, were keen to take a dip.

'Looks like quite a strong current,' I said to Neil. 'Those two people swimming over there, can't get anywhere near the falls.'

Meanwhile, Josh had pulled off his t-shirt off and was wading in.

'Oh, is this really a good idea?' I asked. 'Josh, I think...'

Too late, he had dived under and was now cutting a rapid determined front crawl towards the rocks in front of the falls. He was almost eighteen. Young man enough to make his own decision on whether he was a strong enough swimmer to give it a go.

'He'll be fine. Come on, boys,' said Neil. This was the family I was stuck with (and loved), so I paddled into what turned out to be a miracle hangover cure.

By now Josh had reached the falls and with the water crashing around him he hoisted himself upon a rock and stuck out his arms in a dramatic Christ-the-redeemer pose. From a nearby bridge, Japanese tourists took photos of him. 'Josh might go viral,' Freddie said.

'No, I think the water looks clean,' answered Neil, either misunderstanding what Freddie had meant or cracking a 'dad joke'. It was always so difficult to tell.

Walking back through the forest, we called into a small sanctuary for Asian black moon bears. Caught for their bile and claws, highly prized in Chinese medicine, twenty of these bears had been rescued from the illegal wildlife trade and had undoubtedly led a miserable life until now.

'Have you ever seen a more content looking animal?' I asked, as we watched them napping in their hammocks. One briefly looked our way, then yawned and averted his gaze back up to where sunlight filtered through the mango trees.

'This must seem like heaven after what they've been through,' said Ben. We were all a little mesmerized by their stillness, which embodied exactly what Kharta had told us about Lao culture.

'I'm going to try to be a little bit more moon bear,' I promised.

I felt a kiss on the top of my head from Ben. 'Nah, you're alright, Mum,' he said.

Moments like this were to savour. After all was said and done, Ben didn't really want me to change. Neil and I were like any other parents, at times maddening, embarrassing, interfering and unreasonable in the eyes of our sons, but would they swap us? I didn't think so.

On Sisavangvong Road, chock full of small cosy restaurants and bars twinkling with candles and fairy lights, we settled down at a table in Tangor, a

Lao-French fusion restaurant. Our routine at dinner was to play cards, either Cheat or Old Maid, and I watched Josh deal out our hands. A large tangerine mural of Indochina decorated the wall, scratchy Edith Piaf songs played in the background and an old-fashioned ceiling fan whirred above our heads. It could have been 1940.

'That's nice to see. Simple pleasures,' commented the owner, watching Josh shuffle the remainder of the pack. 'By the way is George bothering you?'

She was referring to a mutt of undetermined breed, who had taken a shine to us, and was leaning against my legs.

'Not a bit,' I told her, giving George a scratch behind the ears. 'Is he yours?'

'No, he's a free spirit, sleeping on the street by night but calling in on friends during the day. He keeps an eye on us all and we pay him in titbits. It's a happy arrangement.'

George appeared to grin at me, peeling back his lips to reveal surprisingly nice teeth. I watched my boy's faces in the candlelight, arguing mildly over what game to play, Like George, they looked happy, by now accepting that Snapchat streaks, Facebook posts and Instagram stories would have to wait. That evening, without me even making the request, their phones had been left at the hotel.

15

Cuba

A Lesson in Freedom

Josh age eighteen, Ben seventeen and Freddie thirteen

15 July 2019,
Parque Nacional Vinales, Cuba

I've still got 'Yeah, I'm gonna take my horse to the old town road. I'm gonna ride 'til I can't no more', going around my head, which was playing from Josh's iPhone as we trotted through a valley of magotes *(nicer than they sound – actually limestone mountains), past fields of tobacco, pineapples and potatoes. When I wasn't shouting* 'mas lento' *to Lucero, my very naughty horse, which was either at a standstill chewing grass or at a trot to catch the other horses up, I was laughing until tears ran. I could hear Neil issuing an ever hopeful 'whoa' every time his broke into a run and, of us, Rafael said that Freddie was the most natural 'gaucho'. As I watched him at a canter some way ahead, I had to resist the urge to spoil the fun by voicing concerns about safety, particularly when we passed another tourist on a horse, coated head to foot in sticky brown mud.*

We stopped at a tobacco farm, where Neil struck a deal with the farmer for twenty cigars, who, referred to us as 'gringos'. Rafael quickly explained that this was a term used by Cubans, simply meaning foreigner. I've watched enough spaghetti Westerns with Dad, to know that 'gringo' is often used disparagingly. Either way, it felt like something that everyone should have on their bucket list. Being called a gringo by a Cuban cowboy. Tick. Meanwhile, Josh was already back on his horse, taking selfies with an unlit cigar in his mouth. 'Both hands! Hold on with both hands!' I screamed as he trotted away, but he couldn't hear me above the track of 'Old Town Road'. What a happy day.

For years I'd dreamed of drinking mojitos in Havana with *son cubano* – a blend of African and Spanish beats – playing in the background. For the boys, it brought the opportunity to ride around in those wide-chassied, tail-finned, iconic American classic cars in rather camp colours ranging from flamingo pink to turquoise blue. With Josh just a month off nineteen, and Ben in his last year at school, I felt this might be our last epic adventure as a family. No doubt that we'd enjoy the occasional week away here and there in the years to come, but, for Josh and Ben, travels with friends and partners would most likely take precedence over family holidays.

'Do you think you'll come away with us next year?'
I asked Josh as we were buckling up for take-off.

'Bloody hell, Mum. We haven't had this holiday
yet,' he said. Truth was, I was already in mourning for
this trip. I had a sinking feeling in the pit of my
stomach that I normally only got on the last day of a
holiday, most likely the first stirrings of panic over
an impending empty nest.

'Well, wherever we go, you're always welcome,'
I offered.

'I know. Thanks,' he told me. 'But next year I might
just do festivals.'

I reminded myself that this was the natural order
of things. At eighteen, I'd have been the same, but in
the meantime, while we were all together, I was
going to make this the best trip we'd ever had. If this
was to be our last adventure, we'd go out on a high:
with no arguments, no mishaps, only fun, fun, fun.
The pressure of it all was giving me a headache.

Havana was just as I'd imagined it. In some parts
of the old city grand colonial mansions were overrun
by nature with trees poking through rooftops; in
other areas, cobbled plazas had recently been
given a make-over for Havana's 500th birthday, and
buildings had been scrubbed of grime, with the
immense Neo-classical domed roof of El Capitolio,
Cuba's seat of government, positively gleaming. I
was in love with the city from the get-go, and much

of that was due to the people. Friendly, noisy, always ready for a chat and to offer advice (even if you weren't looking for it), we soon became used to the question, '*Que pasa*? (what's up), Man,' as we wandered. Josh, Ben and Freddie were high fived by a stream of guys who looked like Nineties hip-hop artists dressed in nylon basketball vests, long baggy shorts and baseball boots, with their caps worn in reverse. I was suspicious of the first few who fell into step with us – unsure if they were going to try and sell us something dodgy or offer to break dance – but it soon became apparent that a chat was all that they were after. 'That your mum? She's cool,' said one particularly nice young man to Josh and Ben, who looked satisfyingly wrong footed.

To explore beyond the city centre, we hired a canary yellow Ford Thunderbird (circa 1955), which came with a driver, dapperly dressed in wide baggy suit pants held up by braces, worn over a crisp cotton shirt. On his head sat a panama hat.

'*Hola*! Slide in,' Jose told us. Neil and Freddie went up front, while Josh, Ben and I shared the bench seat behind. I couldn't remember a time when we'd travelled in such glamorous style. Jose was charismatic and chatty, asking questions about our life in the UK, and Freddie tried out the Spanish he'd recently learned for an oral test at school, which he could only repeat at speed. Some nonsense about his pet rabbit (we don't have one), how on weekends he liked to visit the library (I've never seen Freddie open a book of his own volition) and that his two older brothers were always kind to him (again, debatable).

We drove along the Malecón, Havana's seafront boulevard, where locals stood on the sea wall to fish and young couples on dates strolled past the iconic time-warped Hotel Nacional, which had hosted everyone from Cuban revolutionaries to Hollywood stars. 'Just for tourists now,' Jose told us. 'Too expensive for Cubans.'

Jose's throw away comment hung in the air. Under Communist rule, the wealth divide between Cubans and foreign visitors was massive and although Jose hadn't sounded one bit bitter, I reminded myself that beyond the sunshine, salsa and sparkling sea that the political situation in Cuba was complex, and we needed to be sensitive to that.

Josh was in charge of our itinerary that morning and had chosen to visit a place known as Fusterlandia, which had begun life as a sleepy fishing village called Jaimanitas but had been renamed due to the work of resident artist Jose Fuster, whose two-storey home had been the starting point for a fantastical project in 1975. We pulled into a seemingly modest district. 'Wait for it,' said Jose, as he turned into a side street. It was the eight chimney pots emblazoned with 'Viva Cuba' that caught my eye first. If Gaudi and Picasso had a love child this was the kind of art you might expect them to produce, with the emphasis on 'child'.

'Did an adult do this?' Freddie enquired.

It was true that the mosaic work was naïve, as in simple with an unaffected form, but wow was it uplifting. With themes unmistakably Cuban such as palm trees, cowboys, cactus and couples embracing, Fuster's art was joyous, whimsical and extravagant in scale and colour. Although, I could only imagine what some of his more conservative neighbours must have been muttering when massive sculptural pieces such as a 'cowboy on crutches' were thrown up on his rooftop.

'It's like Gaudi on crack,' commented Josh, taking a photo of a mosaic angel with a smile like the Mona Lisa. Sunburst fountains, psychedelic pools and rippling walkways led to the roof from which we could see that Fuster, not content with decorating his own casa, had strayed out into the community to the pavements, walls and bus stops.

'What did Castro make of it?' I asked Jose. Whatever I thought communism might look like, this certainly wasn't it.

'Oh, he was a fan. Want to visit somewhere else Castro liked?' he asked.

On our drive back into the city, we pulled up to an ice-cream parlour.

I'd read about Coppelia, a state-subsidized

ice-cream business established in the Sixties by Castro and still going strong.

'Try the chocolate. It's the best,' Jose told the boys.

'I don't really imagine Castro eating ice cream,' I said.

'Actually, he had a very sweet tooth. Plus, it was meant to upset the US when the embargo began in '62,' Jose explained. 'And it was a way of keeping the people happy.'

We sat in the sunshine to enjoy this rare sweet treat in a country where you couldn't even buy confectionary ('What nothing at all? Not even Haribo?' Freddie had been aghast),

'Ben, what are you studying?' Jose asked.

Ben gave him a run-down of his A levels, and what he hoped to study at university the following year.

I watched Jose tip back the brim of his panama hat, so as to get a better look at my middle son. 'English?' he repeated, looking puzzled. 'But you are English.'

'English Literature,' Ben explained.

'English books?'

Ben nodded.

'But what for? To be a teacher?' Jose wanted to know.

Josh, who had just finished the first year of an engineering degree, looked delighted at the direction the conversation was going in. 'Yeah so basically, it's a useless degree; he chipped in, purely to wind his brother up.

Ben shrugged. 'Well, because that's what I like.'

We watched as the concept of choosing something to study at university purely on the basis of what you 'liked' (and not to benefit the government of the country that you live in) sunk in.

'Did you go to university?' Ben asked.

'Yes. To study engineering, like Josh.'

'But you don't work as an engineer?'

'Not enough money. I make more as a driver and guide; Jose told us.

Ben smirked at Josh. 'Well, Josh, if you don't enjoy digging drains when you've left uni, at least you have options; he said. Snipes aside, this was our first lesson in what life was like under communist rule in Cuba. Freedom to study at whim wasn't an option.

Travelling on from Havana, we passed through many an idyllic setting, populated by friendly open people, who appeared to be living a pretty good life (full of music, rum and laughter – as the three naturally went hand in hand), with a government that provided excellent free education, healthcare and housing to all, but expected much in return. On the surface, Cubans looked as if they wanted for nothing, but when you talked to them, it was obvious they desired much more.

Wedged into the Sierra de los Órganos mountain range, the town of Viñales is the epicentre of Parque

Nacional Viñales. A strip of pastel-hued casas, small shops and a few restaurants make up the high street, where cowboys tether their steeds to posts and, by night, locals gather to dance outside the community centre under the fire red blooms of Flame trees.

'Time for a Diet-Coke break,' I sighed, as I watched a gum-booted hunk ride by.[26]

'But you're drinking a mojito,' pointed out Freddie, fortunately not understanding my saucy quip.

'Indeed, I am,' I said, taking a sip of my third (by now I'd realized that a mojito wasn't really considered an alcoholic beverage and was fine to drink in quantity at any hour). I was sitting in a swinging chair on the small veranda of our chalet watching fire-flies flit in a dusky sky. Minutes before, we'd watched the sun set over the valley below.

'It's a bit grubby,' commented Neil. For a moment I thought he was referring to my imagination, then realized that he'd run the bathroom tap and a murky brown water had trickled out.

'But the view is five-star,' I reminded him.

This was true. We hadn't travelled here for power showers and good Wi-Fi connection. You could get those at home. We'd come for the scenery and the chance to explore it with a real-life cowboy, and ours

---

[26] The 'Diet-Coke break' advertising campaign ran from 1994 to 2013. Each advert centres around women who stop what they're doing to ogle a builder, window-cleaner ... or in my case ... cowboy.

was called Rafael, who was born in the valley and was third generation cowboy farmer.

Next day, standing on a red-dirt track, we watched as Rafael decided which of his horses to assign us.

'This your horse,' he said to Neil, handing him the reigns. Neil at 6 ft 2' was clearly the tallest among us. His horse on the other hand was not. We watched him step lightly up to sit on top of it.

'You look ridiculous riding that,' Josh said, with his usual honesty.

'Are you sure this one is right for me. Maybe it would be better for Kate?' Neil asked.

'Small but feisty,' explained Rafael, and with a wink and a slap on his bottom (the horse's, not Neil's), the horse took off, leaving the boys and me hooting with laughter.

In the rare moment when our horses were behaving themselves, Rafael chatted to us about the difference between American and Cuban cowboys. The massive one being that, post revolution, cowboys in Cuba were more likely to grow crops than tend to cattle. The much more minor one being that leather cowboy boots weren't worn (since the cattle industry dwindled, leather was a luxury few could afford), but rather plastic gum boots with spurs. To be honest, it was quite a hard look to carry off.

'Ever been to a rodeo in America?' Freddie asked him.

Rafael smiled and shook his head. It's a mind-boggling fact that Cubans aren't allowed to leave the country, unless they're doctors on exchange programmes organized by the government, diplomats or politicians. The reality of life in Cuba hit us like a ton of bricks again as Rafael explained the restrictions he lived under. Thank God for the sight of Neil sitting on his tiny horse to lighten the mood, which despite its size stopped for a gargantuan dump every couple of minutes.

As our travels continued, we spent our money directly with local people, buying cigars straight from a farmer in Viñales, knowing that he'd make more in one transaction with us than several involving the authorities; in Trinidad we stayed in a *casa particulare* (a private home), leaving a large tip for the hosts, who we knew would be paying most of their rental income in taxes; we sought out private guides and whenever possible gave cash in hand. It felt good to know that our money would go solely to the people who had looked after us so well.

After all these years, we still stuck to our formula for a successful family holiday, namely culture first, beach second and so headed last to Varadero, a 12 mile stretch of the whitest softest sand; ranked in the world's top five beaches.

'Good choice, Mum,' said Ben, settling down on his sunbed.

Nice though it was to be appreciated, I was still in shock at being tagged with a fluorescent yellow band on arrival. 'Is it really necessary?' I asked, while the receptionist secured it tightly round my wrist. It wasn't snobbery; I'd never been good with conformity. I left Brownies, aged six, because of the stupid uniform.[27]

'Isn't everyone on full board here, anyway?' I persisted. 'Why do we even need one?'

'You're being embarrassing', Josh pointed out. On a student grant, he was relishing the idea of all-inclusive. His yellow wristband was a ticket to free booze and he clearly didn't want me screwing up the arrangement.

I tried to focus on the positives of the tag, such as the boys not constantly pestering us for drinks, ice creams, kayak-hire or in Josh's case, beer, but couldn't help worrying that the irony of living in a pseudo-world of bottomless glasses and all-you-can-eat buffets in, of all places, Cuba would be lost on them.

'Isn't that Freddie with a donut/ice cream/coke/cake/burger/mocktail/more ice cream?' I asked Neil at regular intervals, as I caught sight of our youngest sauntering around the resort and relishing his wristband autonomy.

I needn't have fretted though as reality checks

---

[27] Brownies is the section in the Girl Guides organization for seven to ten-year-olds, established in Britain in 1914 by Lord Baden-Powell.

were frequent, beginning on our way back from breakfast on the very first morning. Two of the resort's gardeners, one fat, the other thin, sidled up to us sideways in the style of the Chuckle Brothers.[28]

'*Hola, Senora*,' one said cheerily. '*Hola, chicos.*'

'*Hola*,' we all chorused.

'Very nice room you have, *si*?'

'Lovely, thank you.'

'Nice view?'

'Yes, very nice.'

'Nice bathroom?'

'Yes. Lovely bathroom.' I answered, wondering how long this might go on.

'Good mini-bar?'

Ah, okay, I thought, now we're getting to it.

'And with a nice bottle of rum?'

'Yes.'

'The best rum. With the gold label, *si*? Much money for me.'

The boys had cottoned on, too. Josh, who had been apoplectic with happiness on discovering a whole bottle of rum in his room, narrowed his eyes.

'And beer?' the gardener went on.

'Also, beer. Yes.'

'When you drink it, they bring more for you?'

---

[28] The Chuckle Brothers were an English comedy double act active from 1967 to 2018.

'I believe so,' I replied.

'If you don't drink it, we would like it very much.'

'Freddie, run and get the bottle of rum from my room.' Josh, I decided, could keep his.

And so, the rum-running began, all agreeing that it fitted rather nicely to the ideal of Cuba's communist values, that all should be shared equally. We became on first name terms with Roberto (the chubby one), who loitered expectantly each morning and word went quickly around the resort about what a lovely (gullible) family we were. Freddie enjoyed his role as covert courier, scurrying up to our rooms after breakfast to collect the booze, which we left hidden in the shrubs for Roberto to collect while pruning.

I was glad that the reality of life in Cuba was never far below the surface of what could so easily look like any other Caribbean holiday, and was pleased to see our sons' awareness growing daily to the small acts of cunning deployed by staff to make ends meet on what was sure to be a paltry pay. Tapping guests for mini-bar rum was common practice, as was making a few extra dollars for providing special treatment (the best table/sun-lounger/fill-in-the-gap-here), but getting friendly enough with guests to enjoy the occasional free drink or lunch was an art form, expertly deployed by the younger members of staff. After a couple of days, Josh was a bon vivant host, regularly supplying his new friends on the recreation team with rum cocktails.

To be honest I'm not a huge fan of big resort type hotels, and I'm not much for joining in with activities. Give me a good book and the shade of a palm and you'll not hear a peep, but Josh, ever sociable and looking for fun, occasionally got me into situations that took me out of my comfort zone.

'Come on, Mum. Be the fun parent,' he said. Oh, he had me at fun. This holiday was supposed to be the MOST FUN we'd ever had, which was how I ended up trying to balance on a space-hopper like inflatable during organized 'pool games' one day.

We'd been adopted by Carlos and his large noisy Cuban family from Miami (oh, we love your accents. You are both so cute. Be on our team, why don't cha?). I liked them and, ever with my nosy-parker journalist's hat on, I was hoping they might tell me a bit about what it was like for them returning to a country that their parents had fled from post revolution.

'Did you have a good time visiting your family?' I asked Carlos.

'Lemme tell you,' he boomed. 'We just got here from Havana, right? You're not gunna believe what happened, right? You're gunna say I'm making it up! But it ain't that way, honey. Everything I'm gunna tell you is true. So, listen up. Shots!'

A line of shots arrived, and we knocked them

back; my doubts over Carlos sharing his story now disappearing as fast as the rum.

'You been to Havana?'

I nodded.

'So, you'll know where my cousin's place is. Old house, just off Plaza de la Catedral. Shots!'

The bartender lined them up again, and I realized that I would most likely be drunk by the end of Carlos' story.

'So, it's like three o'clock in the morning, and there's this banging. And I say to Louisa, you hear that? She says, 'sounds like someone's knockin' on the door'. Shots!'

'Who was it?' asked Josh.

I slid my shot glass towards Josh. He'd had a whole first year of uni to get used to this kind of drinking.

'So, we jump out of bed and guess who's charging up the stairs? Cops. Three of them. They're shouting, 'Everyone get up. Get your papers.' Now, I know not to mess with these guys, so we do as we're told. We've nothing to hide, right? They take a long time looking at our passports, even though there's nothing to see, and I know what they really want.'

Carlos rubbed his fingers together.

'Money?' suggested Josh.

'Damn right, money. So, I pay up and think that'll be it, but they tell us to move on.'

'Can they do that?' I asked.

'They can do what they like. They say we have until morning to get out. We done nothing wrong, but they could make trouble for us.'

'And for your family I suppose?' I asked.

Carlos nodded. 'Even so, my heart will always be in Cuba,' he said with a shrug. 'And hey, honey, don't look so sad. We're on holiday, right? Shots!'

Back at the beach, I was having difficulty speaking so listened as Josh filled the others in on how 'rubbish' I'd been at pool games and what a 'light weight' I was doing shots. Through slightly blurred visions I gazed at our three sons. Whether studying English literature or engineering (and who knows what path Freddie might yet wander down, perhaps that of a cowboy?), all would be at liberty to choose their own destiny. Cuba had been a window on to another life, and I felt sure that the freedom they enjoyed as young men in the UK would be all the more appreciated after visiting.

A day later, as the plane taxied down the runway at Havana Airport, I scrolled through the photos on my camera, already feeling hopelessly nostalgic for the time we'd enjoyed together.

'Anyway, where are you thinking of going next year,' asked Josh, picking up on our conversation of three weeks ago.

'Costa Rica,' I told him.

Josh grinned. And we both left those two lovely words, full of possibilities, hang in the air.

Epilogue

'Remember that day when you and Dad had food poisoning?' Ben asked me one day during the first Covid-19 lockdown of 2020. 'That was such a great day.'

That wasn't quite my memory of it, as he was referring to a holiday on the Thai island of Koh Samui, when after eating blue swimmer crab Neil and I spent twenty-four horrendous hours hurling up our guts.

'You mean the day you ran completely feral?' I asked.

'You tried to send us to kids' club, but we escaped,' Freddie said. 'And we lay on the sun loungers playing on our DSs all day and ordered pizza from room-service,' chipped in Josh.

I still have the note that the five-year-old Freddie wrote to me that day to let me know that the cleaners had arrived to clean their adjoining room. It says, 'peapol ar in me rom. Luv Freddie x'. I took comfort in the fact that at least he could spell his own name.

'What place are you dreaming of right now?' I asked them.

'Somewhere warm with a beach, like Thailand,' said Josh.

'A European city, perhaps Amsterdam or Paris,' replied Ben.

Immediately I could imagine them. Josh with a backpack; Ben inter-railing with friends. How wonderful, I thought, to have all those adventures in front of you.

'If they're not coming with us on holiday, can we fly business class from now on?' asked Freddie.

When Covid-19 struck the world, locked down together at home, we often reminisced about our travels. 'Remember the time when we...?' someone would begin, and we'd all soon be declaring our version of events, laughing until tears ran. On every trip we'd taken I'd kept journals, which I now dug out to read and amid the funny anecdotes and specifics of these many holidays, I was struck by how many life lessons the boys had been given (for the above episode, it would have been 'seize the day') and the idea for this book came to me. It was a joy to write about happier times, during the months when life on this beautiful planet appeared so bleak.

Looking back, those years had gone by in a blink. The time we'd spent together – often weeks at a time just the five of us – had been the happiest days of my life, but I had to accept that life must move on. Once the boys had flown the nest, Neil and I would take the narrow-gauge railway to Shimla in India; go on a rainforest adventure in Brazil; island hop in the Philippines; and see the aurora borealis in Finland. Oh, I have such plans. And then just when I think I've got it all worked out...

'About Costa Rica, what dates do you think you'll be going?' Josh asked me, during the third national lockdown in February 2021.

EPILOGUE

'Remember that I want to come too,' Ben reminded me.

Ha! So, I hadn't quite seen the last of them yet.

This book tells stories of travelling to far-flung places, but you don't have to roam so far. The importance of spending quality time as a family is perhaps the biggest life lesson of all. I hope my stories raise a smile, a nod of recognition, and give some inspiration for future adventures.

Above all, I wish you the happiest of family travels.

Kate xx

~~~

Acknowledgements

Thank you, Neil. Best partner in life and adventure.

Thank you, Josh. Thank you, Ben. Thank you, Freddie.

This is your story. Thank you for allowing me to break the rule of what goes on tour, stays on tour. I could burst with how much love I have for you all and couldn't be a prouder mum.

Thanks to my literary agent, Darryl Samaraweera, for your gentle (yet determined) way of going about things and for your unfailing courtesy and sense of fun. And to everyone at Artellus, a real family of a literary agency, who I know have been cheering me on.

Thanks to Katie Bond, my publisher at Aurum, who has laughed in all the right places and made astute editorial suggestions. You have championed this book with the kind of enthusiasm all writers dream of. Thanks to Jennifer Barr, Nayima Ali,

ACKNOWLEDGEMENTS

Liz Somers and all at the Quarto Book Group.
Thanks to designer Hannah Naughton for creating
such a beautifully exuberant book cover.

Thanks to Mum, to Dad and my sister, Dode, and
also to Sandra. Your support means the world to me.
Thanks to Rob, who lives on in our memories.

Thanks to all of the local guides and drivers who
we've met on our travels. We've learned so much
from you.

Thanks to Bertie, my westie, for keeping my feet
warm while I write.